Mathematics and Common Sense

Mathematics and Common Sense

A Case of Creative Tension

Philip J. Davis

A K Peters, Ltd.
Wellesley, Massachusetts

Editorial, Sales, and Customer Service Office
A K Peters, Ltd.
888 Worcester Street, Suite 230
Wellesley, MA 02482
www.akpeters.com

Library of Congress Cataloging-in-Publication Data
Davis, Philip J., 1923-
  Mathematics and common sense : a case of creative tension / Philip J. Davis.
    p. cm.
  Includes bibliographical references.
  ISBN-13: 978-1-56881-270-0 (alk. paper)
  ISBN-10: 1-56881-270-1 (alk. paper)
    1. Logic, Symbolic and mathematical--Methodology.   2. Reasoning.
  3. Common sense. 4. Knowledge, Theory of.   I. Title.
  QA9.2D38 2006
  511.3--dc22

                                                        2006012622

Cover images:
First marigold: See pages 4 and 156
Second marigold: Photo by Matt Fitt, www.mattfitt.com
Bee: Photo by Paul F. Worsham

Set in Times New Roman PS MT and Bell Gothic

Printed in India

10 09 08 07 06   10 9 8 7 6 5 4 3 2 1

Contents

Preface

The impetus for putting together this book came from a number of questions about mathematics asked of me by a woman I shall call Christina. Christina had the usual mathematical education of students who were not studying science and technology. She married a mathematician and, over the years, developed an increasing desire to know something more about mathematics. She pointed out to me that what she learned from the media left her with an image of the subject as an arcane, esoteric discipline created by unworldly, mentally disturbed geniuses totally lacking in social skills. The media—newspapers, TV, plays, science fiction, etc.—all deal with what might be called the sensational aspect of the subject: famous problems solved after centuries of work, a new prime number, of almost eight million digits, discovered. She felt, and I agree with her, that these aspects of mathematics, while undoubtedly interesting, did not in any way describe the subject as it was pursued by the average professional mathematician, nor how what they did affected our lives.

Christina asked me a number of questions about mathematics. I answered them, and she was pleased to the point of sending me another batch. Her questions and my responses to them make up the first part of this book. The second part is an elaboration of them.

From the age of about ten, I have studied mathematics, learning its history, applying it, creating new mathematics, teaching it, and writing about it. This has left me with a personal and somewhat idiosyncratic view of the subject (or so my colleagues tell me). I would like to share a bit of this view with my readers.

Mathematics is a subject that is one of the finest, most profound intellectual creations of humans, a subject full of splendid architectures of thought. It is a subject that is also full of surprises and paradoxes. Mathematics is said to be nothing more than organized common sense, but the actuality is more complex. As I see it, mathematics and its applications live between common sense and the irrelevance of common sense, between what is possible and what is impossible, between what is intuitive and what is counterintuitive, between the obvious and the esoteric. The tension that exists between these pairs of opposites, between the elements of mathematics that are stable and those that are in flux, is a source of creative strength.

I go on to other paradoxes. Mathematics is a subject based on logic and yet formal logic is not a requirement for a higher degree. It is thought to be precise and objective with no whiff of subjectivity entering to compromise its purity, and yet it is full of ambiguities. It is the "language that nature speaks" but neither scientists nor philosophers have as yet provided cogent reasons for this.

Mathematics is also an attitude and a language that we employ by fiat and in increasing amounts to give order to our social, economic and political lives. It is a language, a method, an attitude that has diffused into medicine, cognitive science, war, entertainment, art, law, sports; which has created schools of philosophy, which has given support to views of cosmology, mysticism, and theology. All this adds up to a spectacular performance for a subject that in elementary grades has amounted to a rigid set of rules that say "do this and do that."

This book is addressed to all who are curious about the nature of mathematics and its role in society. It is neither a textbook nor a specialists' book. It consists of a number of loosely linked essays that may be read independently and for which I have tried to provide a leitmotif by throwing light on the relationship between mathematics and common sense. In these essays I hope to foster a critical attitude towards both the existence of common sense in mathematics and the ambiguous role that it can play.

These essays show how common sense bolsters or creates conflicts within the ideas and constructions of pure mathematics. Among other things, the reader will find discussions of the nature of logic and the uses of inconsistency, discussions of numbers and what to do with them, of conceptions of space, of mathematical intuition and creativity, of what constitutes mathematical proof or evidence. I will also discuss how common sense operates when mathematics together with its operations engages a wide variety of problems posed by the world of objects of people and events.

Writer Anne Fadiman tells how consternation arose when she tried to amalgamate her substantial collection of books with her husband's equally substantial collection. Each had a "common sense" system of filing; but the systems were mutually incompatible. Later, both Anne and her husband were in minor shock when they heard that after their friend's apartment had been redecorated, the decorator, exercising his own common sense, restored all the books to the shelves according to size and color.

There are more than sixty major subjects or branches of mathematics many of which have significant connections to the other branches. There is therefore no way that the mathematical corpus can be put into a linear order that is totally sensible and consistent. Though the present book alludes to only a very few of these subjects, its chapters may display a certain helter-skelter quality in their arrangement and a slight redundancy in their content. I ask for my readers' indulgence.

I hope that the readers of this book will come out with an appreciation of the flights of human imagination that both join and transcend common sense and that have created the mathematical world we live in.

The Further Reading sections in this book contain references to material that is both popular and professional.

Further Reading

Anne Fadiman. *Ex Libris: Confessions of a Common Reader.* Farrar, Straus, Giroux, 1998.

Acknowledgments

I wish to acknowledge the advice, assistance, information, inspiration, and encouragement given to me by the following people: Carolyn Artin, Thomas Banchoff, Christa Binder, Bernhelm Booss-Bavnbek, Sussi Booss-Bavnbek, Gail Corbett, Chandler Davis, Hadassah F. Davis, Joseph M. Davis, Marguerite Dorian, Amy Drucker, Helaman Ferguson, Michael O. Finkelstein, Julie Gainsburg, Bonnie Gold, Ivor Grattan-Guinness, Charlotte Henderson, Reuben Hersh, Christine Keitel, Laura Leddy, Brandon Lavor, Peter Lax, Yvon Maday, Elena Marchisotto, Stephen Maurer, Carol Mehne, Jan Mycielski, Cleve Moler, David Mumford, Katalin Munkacsy, Igor Najfeld, Kay L. O'Halloran, David Park, Alice Peters, Klaus Peters, Kim Pfloker, Franco Preparata, Elizabeth Roberts, Joan Richards, the late Gian-Carlo Rota, Erica Schultz, Kathleen Shannon, Roger Simons, Ann Spalter, Alice Slotsky, Megan Staples, Charles Strauss, George Rubin Thomas, Peter Wegner, Margaret Wertheim, Roselyn Winterbottom, Alvin White, and Jie Zhang.

Particular thanks go to Ernest S. Davis, who has counseled me throughout the writing of this book. Thanks go also to Gail Corbett, my editor at *SIAM News* at the Society for Industrial and Applied Mathematics. Over the years, she has set challenges for me and given me encouragement. She has also arranged for permission from SIAM to make adaptations from a number of my articles, listed below.

"Mathematics and Common Sense: What Is Their Relationship?" *SIAM News*, Vol. 28, No. 3, November 1995.

"Mickey Flies the Stealth." *SIAM News*, Vol. 31, No. 3, April 1998.

"Astrology: Early Applied Mathematics?" *SIAM News*, Vol. 33, No. 5, June 2000.

"Playing Ball Probabilistically." *SIAM News*, Vol. 35, No. 3, April 2002.

"Mumbo Math." *SIAM News*, Vol. 37, No. 9, November 2004.

"When Mathematics Says No." In *No Way: The Nature of the Impossible.* Philip J. Davis and David Park, eds., pp. 172 ff. W. H. Freeman, 1987.

"Paraconsistent Thoughts about Consistency." *Mathematical Intelligencer*, Vol. 24, No. 2, Spring 2002.

"Letter to Christine." *Newsletter, European Mathematical Society,* March 2004. Also in *Matilde: Nyhedsbrevfor Dansk Matematisk Forening,* Nummer 20, Juni 2004.

"Applied Mathematics as Social Contract." *Zentralblatt für Didaktik der Mathematik,* 88/I. Also in *The Mathematics Magazine,* Vol. 61, No. 3, June 1988, pp. 139–147.

"When Is a Problem Solved?" In *Current Issues in the Philosophy of Mathematics From the Viewpoint of Mathematicians and Teachers of Mathematics,* Bonnie Gold and Roger Simons, eds. Mathematical Association of America, 2006.

To
Christa and Peter
Christine and Fritz
and
Kay
in friendship

We are enlarged by what estranges.

—Richard Wilbur, "A Wall in the Woods: Cummington"

This is the vision ... to endure ambiguity in the movement of truth and to make light shine through it.

—Jonah Gerondi (c. 1200–1263), *The Gates of Repentance*

Letters to Christina: Answers to Frequently Asked Questions

The Swiss mathematician Leonhard Euler wrote a highly admired three-volume popularization of science entitled *Letters to a Princess of Germany* (published: 1768–1772). Some time ago, my friend Christina (name changed) sent me a letter with a number of questions about mathematics, a subject about which she had heard much but knew little. I believe that her questions are among those frequently asked by the general public.

Christina's questions and my answers provide an overview of the mathematical content of this book and set the stage for the essays that follow.

Q1. What is mathematics?

A1. Mathematics is the science, the art, and a language of quantity, space, and pattern. Its materials are organized into logically deductive and very often computational structures. Its ideas are abstracted, generalized, and applied to topics outside mathematics.

The mix of mathematics and outside topics is called *applied mathematics*. Mathematics has often been called the "handmaiden of the sciences," because of its use in and interactions with the other sciences. For example, for reasons that are by no means clear, mathematics is an indispensable aid to the physical sciences. The expositions of theoretical physics are completely mathematical in character.

Applied mathematics includes *descriptions, predictions,* and *prescriptions.* Description replaces a real-world phenomenon with a mathematical surrogate: A lampshade casts a parabolic shadow on the wall. Prediction makes a statement about future events: A total eclipse of the sun will occur on July 11, 2010; its duration will be 5 minutes and 20 seconds. Prescription (or formatting) organizes our lives and actions along certain lines: Traffic lights control the flow of automobiles in a periodic fashion; tax laws affect us all.

Mathematics has intimate relations with philosophy, the arts, language, and semiotics (the theory of signs and their use). Since the development of mathematics is often inspired and guided by aesthetic considerations, mathematics can be described as "amphibious": It is both a science and one of the humanities.

Q2. Why is mathematics difficult, and why do I spontaneously react negatively when I hear the word?

A2. There are many reasons why the average person finds mathematics difficult. Some of them are poor, uninformed teaching; over-concentration on the deductive aspect of the subject; boring presentation; presentation that fails to connect mathematics with the day-to-day concerns of average people.

Mathematical thinking and manipulations are cerebral activities that, simply, not everyone enjoys. And then, let's face it, the material can be difficult, common sense seems to be irrelevant. While mathematical skills and understanding can be learned and developed, I believe there is such a thing as innate talent for mathematics. Just as not everyone has the talent to create art, write a great book, be a ballet dancer, or break athletic records, not everyone can scale the heights of mathematical understanding. The very fact that professional mathematicians make long lists of unsolved problems attests to the fact that ultimately, all mathematicians reach their own limits of mathematical accomplishment.

Q3. Why should I learn mathematics? History widened my horizons and deepened my "roots." When I learned German, it opened up cultural treasures to me. Karl Marx explained (if not changed) the society I live in. What does mathematics have to offer?

A3. I'll start by providing the time-honored reason. Mathematics trains us to think logically and deductively. I would rephrase this by saying that mathematics opens up the *possibility* of rational thought in contrast to unreasoned or irrational, or mystical, thought.

Mathematics is one of the greatest human intellectual accomplishments. We learn mathematics partly to enable us to function in a complex world with some intelligence, and partly to train our minds to be receptive to intellectual ideas and concepts. Even as natural languages such as German or Chinese embody cultural treasures, so does mathematics: One has to learn the "language" of mathematics to appreciate its treasures. To learn the history of mathematics is to appreciate its growth within the history of general ideas.

Looking about us, if we are perceptive, we can see not only the natural world of rocks, trees, and animals, but also the world of human artifacts and human ideas. The ideas of Karl Marx (and of other thinkers) explain the world along certain lines, each necessarily limited by the thinker's own constraints. Mathematics explains the world in remarkable but limited ways. It is capable of formatting our lives in useful ways, for example, when we take a number for our "next" at the deli, and of allowing us to look into the future and make prudential judgments; for example, the meteorological predictions based on mathematics theories. Every educated person should achieve some appreciation of the historical role that mathematics plays in civilization in order to give the subject both intelligent support and—when it seems necessary—intelligent resistance to its products.

Q4. How has mathematics changed in the last 100 years? What have been the dominant trends?

A4. The number of new mathematical theories produced since 1900 is enormous. During that time, mathematics has become more abstract and deeper than it was in the 19th century. By this I do not mean that simple things such as arithmetic have to be viewed by the general public in a more abstract or deeper way than before, but that the conceptualization of old mathematics and the creation and applications of new mathematics by mathematicians have had that character.

Mathematical logic is now firmly on the scene, as is set theory. A new and more abstract algebra has grown mightily and consolidated itself. Infinitary mathematics (i.e., the calculus and its elaborations) has grown by leaps and bounds, but recently has had increasingly to share the stage with finitary mathematics.

Geometry, which began in antiquity as visual and numerical (lengths, areas, volumes), moved into an abstract axiomatic/deductive mode and now includes such topics as algebraic, combinatorial, and probabilistic geometry.

A major trend since the late 1930s is to view mathematics as the study of deductive structures. Mathematics has shared a structuralist point of view with numerous non-mathematical disciplines ranging from linguistics to anthropology to literary criticism.

There is no doubt that the electronic digital computer that emerged in the middle 1940s, which is in many ways a mathematical instrument, has changed our day-to-day life noticeably. The computer has changed mathematical education. It has revitalized or widened the scope and increased the power of a number of traditional branches of mathematics. It has also created new branches of the subject and, all in all, has had a revolutionary effect throughout science and technology.

Q5. What can you tell me about the "chipification" of mathematics?

A5. A large part of the mathematics that affects our day-to-day lives is now performed automatically. Computer chips are built into our wristwatches, automobiles, cash registers, ATMs, coffee machines, medical equipment, civil-engineering equipment, word processors, electronic games, telephone and media equipment, military equipment, IDs... The list is endless, and the idea of inserting chips to do clever things has become commonplace.

The mathematics in these chips is hidden from view. The average person, though greatly affected by it, needs to pay no conscious attention to its mathematics. That is the job of the corps of experts who design, implement, monitor, repair, and improve such systems. The "chipification of the world" is going forward at a dizzying rate. One of the disturbing side effects of this trend may very well be that in the near future our sons and daughters will not learn to add or read in the conventional way. Will this put us back to several hundred years ago, when literacy and numeracy were relatively rare achievements? Not at all. Modes of communication, interpretation, and social arrangements will change; it does not mean that they will disappear.

Q6. Where are the centers that develop new mathematics? Who is "hip"? A fan of good rock music knows where to go to hear it. Where would a math fan go these days?

A6. There are mathematical centers in all the developed countries of the world. They are located in universities; governmental agencies; research hospitals; scientific, economic or social "think tanks"; and industries. All these centers tend to be organized into groups and tend to concentrate their research on very specific problems or subject matter. There are also many mathematical researchers who work independently and productively outside of groups.

Depending on what sort of "mathematical music" you would like to hear, you would visit one of these groups or individuals. Although the groups exhibit a great deal of *esprit de corps* and self-esteem and can be influential beyond their own walls, there is, in my judgment, no single group that is predominates; this diversity is a very good thing. Closer to home, there are now websites that cater to every level of mathematical interest and engagement.

Q7. How is mathematics research organized? Who is doing it, who is paying for it, and why? Lonely, harmless "riders" or highly efficient, highly organized, secret, threatening groups? Should one be scared?

A7. The forward movement of mathematics is driven by two principal forces: forces on the inside of the subject and forces on the outside.

Forces on the inside perceive certain questions or aspects that are incomplete, unanswered, and call for answers. New mathematical ideas arising from free mental play can come into prominence. "Lonely riders" very often make contributions.

Forces on the outside call for the application of available mathematics to the outside world of people or things. These applications may result in the development of genuinely new mathematics.

The work is paid for by institutions (including universities) or organizations in anticipation of profit, or of scientific, social, national, or cultural gain. The economically independent mathematical researcher exists but is quite rare. The space agency NASA is, for example, one source of considerable mathematical support.

In the years since 1939, and particularly in the United States, a great deal of mathematical research has been paid for by the military. Individual mathematicians have profited from this support independent of their personal views of the United States military or its foreign policy. Since war appears to be endemic to the human situation, this support will undoubtedly continue.

Other support comes from the medical and healthcare sector, and this support is likely to increase in importance. The productivity of the United States gets a considerable boost from the entertainment industry, and today's movies use computer graphics software with a considerable mathematical underlay.

Some of the work of the various groups is restricted or confidential. This may be for reasons of national security or for reasons of industrial confidentiality in a free and competitive market.

In an ideal democratic society, all work must ultimately come into the open so that it may be judged both for its internal operation and its effects on society. Free and complete availability of information is one of the hallmarks of an ideal science. While this has not always been the case in practice, judging from the past several centuries, the record is pretty good. Doing mathematics, developing new mathematics, is simply one type of human activity among thousands of activities. As long as constant scrutiny, judgment, and dissent flourish, fear can be reduced.

Q8. Other sciences have had breakthroughs in the 1980s and 1990s. What breakthroughs does mathematics claim?

A8. We might begin by discussing just what is meant by the phrase "a breakthrough in mathematics." If we equate breakthroughs with prizes, and this equivalence has a certain merit, then the work of, e.g., the Fields Medal winners should be cited. This would cover pure mathematics, where this prize, given since 1936, is often said to be the mathematical equivalent of the Nobel Prize. Similar prizes of substantial value exist in various pure mathematical specialties, applied mathematics, statistics, computer science, etc.

The work of Fields Medal winners over the past decades, people such as K. F. Roth, R. Thom, M. Atiyah, P. J. Cohen, A. Grothendieck,

S. Smale, A. Baker, K. Hironaka, S. P. Novikoff, and J. Thompson, cannot, with some very few exceptions, be easily explained in lay terms. In view of their arcane nature, the public rarely hears about such things. Should a particular accomplishment reach the front page of the newspapers, its meaning is often mutilated by packaging it in the silver paper of sensation. Sensation is always easier for the public to grasp than substance.

Judging from the selection of Fields Laureates, the criteria for the honor seem to be (1) the solution of old and difficult mathematical problems, (2) the unification of several mathematical fields through the discovery of cross connections and of new conceptualizations, and/or (3) new internal developments.

A few years ago the world of research mathematicians was electrified by the public announcement that the Clay Mathematics Institute, a private organization, was offering $1 million each for the solutions of seven famous mathematical problems.

We live in an age of sensation. As a result, only the sensational aspects of mathematics get much space in the newspapers: the solution to Fermat's "Last Theorem," "the greatest prime number now known is...," the solutions of other so-called "big" unsolved problems, international students' contests, etc.

Mathematical applications that are useful, e.g., the programming and chipification of medical diagnostic equipment, are rarely considered to be breakthroughs and make the front pages. Historic research or philosophic discussions of the influence of mathematics on society have yet to be honored in prestigious manners.

From the point of view of societal impact, the major mathematical breakthrough since the end of World War II is the digital computer in all its ramifications. This breakthrough involved a combination of mathematics and electronic technology and is not the brainchild of a single person; hundreds, if not hundred of thousands, of people contributed and still contribute to it.

Q9. Medical doctors fight cancer, AIDS, and SARS. What is now the greatest challenge to modern mathematics?

A9. One might distinguish between "internal" problems and "external" problems. The former are problems that are suggested by the operation of the mathematical disciplines themselves. The latter are problems that come to it from outside applications, e.g., what airplane shape has the minimum drag when the airplane surface is subject to certain geometrical conditions? What are the aerodynamic loads on the plane structure during maneuvers?

In 1900, the mathematician David Hilbert proposed a number of very significant unsolved problems internal to mathematics. Hilbert's reputation and influence was so great that these problems have been worked on steadily, and most of them have been solved. Setting up, as it did, a hierarchy of values as to what was important (every mathematician creates his own list of unsolved problems!), this list has had a considerable influence on the subsequent progress of mathematics. The solvers, in turn, have gained reputations for themselves in the mathematical community.

Within any specific field of mathematics, the practitioners will gladly tell you what they think the major unsolved problem (or challenge) is. Thus, if a topologist is queried, the answer probably is to prove the Poincaré Conjecture in the case $n = 3$. If an analyst is queried, the answer might be "Prove the Riemann Hypothesis."

As for external problems, a fluid dynamicist might say "Devise satisfactory numerical methods for processes [such as occur in turbulence or in meteorology] that develop over long periods of time." A programming theorist might say "Devise a satisfactory theory and economic practice for parallel computation."

If your question is answered in terms of specific problems, it is clear that there are many of them, and there is no agreement on how to prioritize them. Researchers can be drawn to specific problems by the desire for fame, or money, or simply because their past work suggests fruitful approaches to unsolved problems.

Your question can also be answered at a higher level of generality. A pilot assessment of the mathematical sciences prepared for the United States House (of Representatives) Committee on Science, Space, and Technology, identifies five interrelated long-term goals for the mathematical sciences. These are

- to provide fundamental conceptual and computational tools for science and technology;
- to improve mathematics education;

- to discover and develop new mathematics;
- to facilitate technology transfer and modeling;
- to promote effective use of computers.

[*Notices of the American Mathematical Society*, February 1992]

I would like to go up one more rung on the ladder of generality and answer that the greatest challenge to modern mathematics is to keep demonstrating to society that it merits society's continued support. The long history of mathematics exhibits a variety of mathematical intents. Some of these have been to discover the key to the universe, to discover God's will (thought to be formulated through mathematics), to act as a "handmaiden" to science, to act as a "handmaiden" to commerce and trade, to provide for the defense of the realm, to provide social formats of convenience and comfort, to develop a super brain—an intelligence amplifier of macro proportions.

Mathematics can and has flourished as a harmless amusement for a few happy aficionados both at the amateur and professional levels. But to have a long and significant run, mathematics must demonstrate an intent that engages the public. If the intent is simply to work out more and more private themes and variations of increasing complexity and of increasing unintelligibility to the general public, then its support will be withdrawn.

The public demands something in return for its support, but the place and the form of an acceptable return cannot be specified in advance. Perhaps a mathematical model of brain operation will be devised and will lead to insight and ultimately to the alleviation of mental disease. It is not too fatuous to think that many of the common problems that beset humankind such as AIDS, cancer, hunger, hostility, and envy, might ultimately be aided by mathematical methods and computation. But while efforts in these directions are praiseworthy, they may come to naught.

By way of summary, the greatest challenges to mathematicians are to keep the subject relevant and to make sure that its applications promote human values.

Q10. What can you say about the militarization, centralization, regionalization, and politicization of mathematics?

A10. This is a very wide-ranging question: Militarization alone would require several books. Since about 1940, one of the major financial supporters of mathematical research in the United States has been the military or the "military industrial complex"; a similar statement can be made of all the advanced nations. There are essential mathematical underlays to new, sophisticated weaponry, both offensive and defensive, and to military information processing systems. Related economic, demographic, strategic studies and predictions often involve complex mathematics.

Since the end of the Cold War, with the development of a variety of insurgencies and terrorist strategies, military options have been and are being reassessed, which will undoubtedly lead to new developments in mathematics. We can also foresee a time when the development and application of mathematics will be increasingly supported by fields such as medicine, biology, environment, transportation, finance, etc.

Centralization and regionalization: In the pre-computer days it was said that a mathematician didn't require much in the way of equipment: a few reams of paper, a blackboard, and some penetrating ideas. Today, although much, and perhaps most, research is still done that way, we find increasingly that mathematicians, particularly applied mathematicians, require the aid of supercomputers. This is still far less costly than the billions of dollars of laboratory equipment required by high-energy physicists or astrophysicists.

Centralization of mathematics occurs as the result of a number of factors, including the willingness and ability of a society to support mathematical activity and the desire to have mathematicians work in groups or centers. The notion of a critical intercommunicating mass of creative individuals is at work here. New systems of rapid intercommunication and the transport of graphical and printed material may affect the clumping of future centers for research and development.

Politicization: While mathematical content is abstract, mathematics is created by people and is often applied by people to people. It is to be expected, then, that the creation and application of mathematics should be subject to support, pressure, monitoring, and suppression by governmental, political, or even religious institutions. The interaction between mathematics and human institutions has a long and documented history.

Q11. Give me ten points that worry a concerned mathematician.

A11. A concerned mathematician will worry about the abuse, misuse, or misinterpretation of mathematics or its applications. Insofar as we are living in a thoroughly mathematized civilization, the number of concerns is necessarily vast. Many such concerns focus on "life and death" issues. If, for example, a mathematical criterion were developed via encephalography for determining when a person is brain dead, then this would engender a great deal of concern.

Q12. I read a statement attributed to the famous physicist Max Born that the destructive potential of mathematics is an imminent trend. If that is so, why should I, an average person, learn more about mathematics?

A12. You should learn more about it for precisely that reason.

All creative acts have destructive potential. To live is to be at risk, and no amount of insurance can reduce the risk to zero. Moreover, to live at the very edge of risk is thought by some people to make them feel "truly alive." Electrical outlets in the home are not totally risk-free; the destructive or revolutionary potential of graphical arts, or of litera-ture, is well documented. Even as mathematics solves many problems, it creates new problems, both internal and external.

The more the average person knows about mathematics, the better off that person is to make judgments. Some of those judgments will be about how to temper risk with prudence. In a world in which scientific, technological and social changes occur rapidly, a democratic society cannot long endure in the presence of ignorance.

Q13. What is deep mathematics and what is not?

A13. A quick answer is the one given by logician Hao Wang: A deep theorem is a formula that is short but can be established only by long proofs.

I, however, am going to answer differently than Hao Wang by changing your question just a bit and discussing the possibility that Mathematics with a capital M differs from mathematics with a lower-case m. First, a cautionary quotation:

Insecure intellectuals make a false and basically harmful distinction between "high" and vernacular culture, and then face enormous trouble in trying to determine a status for significant items in between, like Gershwin's *Porgy and Bess* or the best of popular science writings.

— *Crossing Over: Where Art and Science Meet,* Stephen J. Gould and Rosamond Wolff Purcell

Some months ago, I worked out a certain piece of mathematics that gave me much pleasure and that I believed was new and interesting. I thought about building it into a paper and then began to think to what periodical I might appropriately send it. Then I stopped short and said to myself: "You know, there is Mathematics with a big M and mathematics with a small m. What I've done here is of the small m variety. If I send it to periodical *XYZ*, it would be rejected out of hand."

What do I mean by Mathematics with a big M and mathematics with a small m? It would be impossible for me to present a list of criteria, and my criteria would not necessarily be my colleagues' criteria, but as the saying goes: "I know them when I see them." You have most likely heard of Art with a capital A and art with a small a. Possibly you've heard of Opera with a big O (grand Opera: Wagner, Verdi) and opera with a small o (opéra comique: Offenbach, light opera, Broadway musicals). Then there is poetry, verse, light verse, and doggerel. Recent articles have dealt with the movements back and forth within the categories of big and small. The work of Norman Rockwell—a popular American magazine cover artist, once thought to be the Rolls Royce of kitsch—has now been reconsidered and elevated in the minds of art critics.

I will end by adapting a paragraph of the philosopher William James. I have changed a few of James' words so the paragraph relates not to the split of philosophers between the tough- and tender-minded, but to mathematicians.

It suffices for our immediate purpose that the M and m kinds of material, both exist. Each of you probably knows some well-marked example of each type, and you know what the authors of each type think of the authors on the other side of the line. They have low opinions of each other. Their antagonism, whenever as individuals their temperaments have been intense, has formed in all ages a part of the mathematical atmosphere of the time. It forms part of the mathematical atmosphere today.

Despite all the fuzziness and inconsistency (often in my own mind), the dichotomy remains alive in the mathematical world. It can be the source of professional snobbism: "X is not a 'real' mathematician," "Y doesn't prove theorems," "Z only computes." The split infects the way professional talks are given. It can play a role in job offers, promotions, and in obtaining contracts and grants. Though attitudes change, the dichotomy is not likely to go away soon.

Letters to Christina:
Second Round

Christina must have been stimulated by my answers to her questions, because, a few months later, she sent me a second set of questions. I suspect that in the meantime she must have done a bit of independent reading, because to answer her questions now required more thought and elaboration on my part.

Q14. Can mathematicians look at a formula or an expression or an open problem and tell almost immediately whether this is something deep or something easy?

A14. I've often wondered whether a musician or the conductor of an orchestra can look at a new musical score and get a feeling for what the composer has created from the printed page. Along a totally different line, my wife said to me recently, "After all these years of cooking, I think I can now look at a recipe and have an idea how the dish will turn out."

The answer to your question is yes and no and everything in between. Thesis advisors to Ph.D. candidates in mathematics often suggest problems to their students. The suggested problem should be neither too hard nor too easy. In general, the advisor bases his or her recommendations on long experience, knowledge of the mathematics, and a judgment of the capabilities of the student.

When I first went to my thesis advisor for a problem, he suggested one that I was able to solve in a few days. That was the end of that suggestion. Then he suggested a problem that was both difficult and uninteresting for me. And that was the end of that advisor. I looked around—successfully—for a new thesis advisor.

A research mathematician may attempt to solve an open problem that has been around for a while. Unless it happens that no one has else has worked on it, a reasonable assumption is that the problem is fairly difficult.

A research mathematician may very often push forward and make up a problem or a theory of his own. At this stage, he probably has only a vague idea of the depth or the difficulty that the problem presents, or where the problem might lead. He will do what he can with the problem or theory that he has created. Other mathematicians might then come along with more knowledge, technical skills, creative abilities, or just plain luck and push the theory beyond its original conception or context.

Q15. What can you say about independent or "dissident" status of mathematical arguments? There are around us hundreds of paid spin-doctors, thousands of experts, hundreds of thousands of journalists and politicians. They argue for and against many things. Can any single mathematician, on occasion, act like the little boy in the Hans Christian Andersen story of *The Emperor's Clothes*, and say "no way"?

A15. There are many disputes that can be settled by the simple mathematical operations of counting and measuring. A woman decided she would like to move a chest from room A to a place in room B. The husband said it would never fit; the wife insisted it would. The dispute was easily resolved by measuring the widths of the chest and of the space. I wish it were the case that all disputes could be resolved so easily by mathematics!

The great 17th-century mathematician and philosopher Leibniz had law training. He dreamed of a method or a calculus of reasoning that he called the *Characteristica Universalis*. Once humanity was in possession of it, if a dispute arose between individuals, it could be settled easily by a mathematical computation. Well, despite all the developments in mathematical logic since Leibniz's day, the existence of such a calculus seems to be a will-o-the-wisp.

Consider a lawsuit, either civil or criminal. Perhaps the prosecution or the defense has brought in a mathematician to testify as an "expert witness." Very likely the mathematician will produce a statement involving probability or statistics. It is often the case that there are many ways in which the opposing lawyer can "destroy" the testimony of the mathematician or induce doubts in the minds of a jury:

(a) by producing another mathematician who will testify to the contrary;

(b) if a certain event or is highly probable but not absolutely certain, then in the particular case under litigation, the opposing lawyer may raise doubts as to whether it actually happened;

(c) by questioning the theoretical assumptions under which the statement of probability has been derived;

(d) by questioning the accuracy of the data upon which calculations may have been made.

Here is a quotation from the introduction to the book *Statistics for Lawyers* by Michael O. Finkelstein and Bruce Levin:

> Do statistics really matter? This is a question that sometimes vexes statisticians. In the legal setting, the questions are whether statistical models are fairly evaluated in the adversary process and whether statistical findings are given their due in the decisions. Unfortunately, the record here is spotty, even perverse. In some cases the courts have appraised statistical evidence well, but in some important public issue litigations very good statistical models have been summarily rejected and very bad ones critically embraced by judges and Justices in pursuit of their own agendas.
>
> The lawyer of the future predicted by [famous American Supreme Court Justice] Oliver Wendell Holmes, Jr., ninety years ago or indeed of Leibniz three hundred years has not yet come into his or her own.

Let me move away from the law. Politicians, journalists, spin-doctors, and "authorities" often attempt to prove their case by citing numbers. "Do the math" is a common expression meaning that if you "did the math" you would be absolutely convinced. Often, however, contradictory pieces of mathematics are presented in argument. When the arguments are examined carefully, and if the math is bona fide, it is usually the case that the differences result from considering different aspects of the case in question. There is another term that has become a common expression: "fuzzy math." This essentially means any argument at all that comes wrapped in a kind of mathematical verbiage.

Professional socio-psychologists also dispute interpretations and conclusions derived from mathematical arguments. The most notorious instance of this that I can think of arose in the case of intelligence testing leading to numbers that are called IQ's. Another frequent dispute at the legal level is the reliability of genetic information obtained in DNA testing.

In summary, I would say that at the present time in the legal, political, social and psychological arenas, mathematics can be a two-edged sword. It is lucky that there are numerous mathematicians who, like the little boy in Andersen's story, stand up and protest fallacious statements.

The utility of mathematics in settling questions of science and technology has its own characteristics. Not long ago, a very distinguished physicist said to me "Look, all this talk about space travel to the stars— not the planets—is nonsense. It's science fiction. Let me show you a computation that proves it." The possibility of saying "yes" or "no" to a physical statement based on a mathematical argument is a very complex matter and would require several letters to you at the very least.

Q16. What is the power of mathematical formalisms? Does the power come from the ease with which logical transformations of statements into other statements can be made? Or is it that, on the one hand, formalisms support the memory by the encoding of experiences and insights that have already been made, and on the other hand support the imagination and therefore free the mind of fixed prejudices?

A16. Some of the power of formalisms can be understood easily and some of their power remains quite mysterious.

First a word as to what I mean by the word "formalism." I will limit the meaning to lines of text that embody or are written in specialized mathematical notation.

1. The Pythagorean theorem as a formula is $c^2 = a^2 + b^2$.
2. A theorem in first-year calculus states that $\int x dx = x^2/2$.

Both of these lines can be stated in a natural language (say English or Danish) in different ways:

1a. c multiplied by itself equals a multiplied by itself plus b multiplied by itself.

Or, by giving the specific historical and geometric context,

1b. The square on the hypotenuse of a right triangle equals the sum of the squares on the two legs.

Statement 2 might be restated as

2a. The integral of xdx equals x squared divided by 2.

Giving Statement 2 a geometrical context, we might say

2b. The area of an isosceles right triangle is one half the square of the length of its leg.

One power or utility of the mathematical formalism becomes apparent: Statements 1 and 2 are relatively context free. They do not have to be interpreted geometrically; they can be interpreted solely in terms of certain arithmetic or calculus operations that have been agreed upon. Statements 1 and 2 are therefore free to be embedded in other contexts such as 1b or 2b.

Aside from the context-free aspects of mathematical formalisms, what other powerful features do these statements exhibit? They exhibit compactness, preciseness, suggestibility, ease of manipulation, aesthetic qualities, and universality. I'll describe these features one by one.

Compactness. Compactness is easily understood. Statement 1 requires eight characters while Statement 1a requires 61 characters. Statement 2 requires 10 characters while Statement 2a requires 41. Mathematical formalisms therefore act as a kind of shorthand.

Preciseness. Mathematical language, by virtue of its agreed-upon nature, is very likely the most precise of all the languages. The symbols 10.10 and 10.11 mean completely different things. In some contexts, mixing them up (as in the case of phone numbers or IDs) can cause confusion or even damage. In other contexts (as in estimating your checking account balance), no great harm may result. Natural languages allow a "stretch of meaning" or a certain amount of ambiguity. Mathematical language minimizes ambiguity. This is its strength and occasionally its weakness.

Another language with a great amount of preciseness is musical notation. C sharp is C sharp and that's all there is to it. But orchestral conductors in conducting a musical score can arrive at variant interpretations that are distinguishable one from the other. And over the years, orchestral conductors have retuned the scale a bit higher.

Suggestiveness. Here are some examples: At one point in mathematical history, the quantity a^2 was written as aa. Juxtaposition designated multiplication of the quantities juxtaposed; it is still employed (although not in numerical computer languages) when the product of a and b is written as ab. When the same quantity is multiplied over and over as in aaa or $aaaaa$, it was easier to write a^3 or a^5. Having displayed a mix of an unspecified symbol (a) with a number, the question suggests itself: What meaning might be assigned to, e.g., $a^{3.5}$. Of course, the question might have been phrased as "what meaning might be assigned to the operation of multiplying a by itself three and a half times," but the law of exponents, written symbolically as $(a^m)(a^n) = a^{m+n}$ and invoked to give an answer (but not the only possible answer) to the riddle, is much more obvious than the meaningless process of doing something "three and a half times."

The suggestibility of mathematical symbolism is often summed up in the statement that "the symbols are smarter than we are." This almost suggests that there is something magic in symbolism. This idea, which is very old, must be watched very carefully. It is the basis of talismans and a variety of systems of mysticisms.

Ease of manipulation. Consider the algebraic identity $a(b + c) = ab + ac$. If it is necessary to do so, it is much easier for humans to proceed from the left-hand side of the equation to the right-hand side. And it is much easier for humans to program a computer to do this when the principle is written out in algebraic notation than if the principle were embedded in a natural language: "the product of a and the sum of b and c equals the sum of the product of a and b and the product of a and c."

Aesthetic qualities. Somewhat harder to grasp are the aesthetic qualities that formalisms often display. The formula $e^{\pi i} = -1$ is thought by mathematicians to be one of great beauty because it displays four basic constants of mathematics in a compact, surprising, and tremendously useful relationship in that it has many consequences.

Universality. Mathematical notations, symbolisms, or formalisms are part of what may be called the language of mathematics. A mathematical sentence consists partly of symbolisms and partly of natural language. This mixture of natural and symbolic languages often promotes ease of understanding by interlarding the symbols with natural language. As a young researcher in mathematics, one of my professors gave me the following piece of advice for writing mathematical papers: Embed every mathematical statement within a grammatically correct English sentence. Here is an example taken from E. C. Titchmarsh's very influential *The Theory of Functions.*

> Let $f(z)$ be an analytic function, regular in a region D and on its boundary C which we take to be a simple closed contour. If $|f(z)| \leq M$ on C, then $|f(z)| < M$ at all interior points unless $f(z)$ is a constant (when, of course, $|f(z)| = M$ everywhere).

While this paragraph could have been written entirely in mathematical symbols, it would have been both tedious and much more opaque, even to a professional reader. Mathematical language has developed over a long period of time. Many older symbolisms were actually abbreviations of natural language expressions. As new theories emerge, new symbolisms are born. Symbolisms can also die. The history of mathematics contains a graveyard of dead symbolisms.

There appears to be a process at work that may be stated as "form follows function," but the reverse is also true: "we make our language, and then our language makes us." This is called the Sapir–Whorf Hypothesis in semiotics.

In recent years, many computer languages have been developed by individuals or by groups. Many of these languages are now dead or obsolescent, replaced by new languages and new hardware. Many programming languages that were once widely used and studied now continue in at most marginal form, in "legacy code," i.e. code that is no longer supported by the manufacturers. But programming languages tend to survive in a marginal way much longer than one might guess.

Someone who knows no English may not be able to read the English part of a mathematical sentence, but since the symbolic part is universally accepted and understood, he will be able to grasp much of its meaning.

Q17. Why is it that quite different things can look similar in mathematics?

A17. This question is not as innocent as it appears. There are many different answers to it, each emphasizing a different aspect of mathematical experience and usage, and all interrelated. In my opinion, however, these various answers, taken together, do not provide an adequate explanation. But I shall answer as best I can.

I'll give answers

1. from metaphor,
2. from available language,
3. from the unity of nature.

1. Mathematics as metaphor. Mathematical text is embedded in and is an outgrowth of natural languages (e.g., English, Danish, Chinese, etc.). Remove the natural language component and the formal symbolic, formulaic portion of mathematics becomes meaningless. Take formulas such as $\sin^2 + \cos^2 = 1$, $AA^{\mathrm{T}}A = A$, or $e = mc^2$ and try to give meaning to them, using only additional mathematical or logical symbols and without the use of natural language. It's impossible.

A metaphor, according to my American Heritage Dictionary is "a figure of speech in which a term is transferred from the object it ordinarily designates to an object it may designate only by implicit comparison or analogy." Natural languages are full of metaphors; in fact, some students of linguistics say that we cannot communicate with one another except through metaphoric language. Take the word *green,* for example. The initial, and perhaps primary meaning of the word, is a certain color. But green has many metaphorical usages; a boss may "give the green light" to a project; a "green" person is unskilled in certain tasks ("in my salad days, when I was green in judgment," Shakespeare, *Antony and Cleopatra*) there is a Green political party, a green card. The word green is also found as an expression of hope, as the designation of a team of Byzantine charioteers, as a metaphor for fury, envy, money and many other things. The word green has wandered all over the map of human concerns starting from the color of grass and lettuce.

All mathematics can be considered to be metaphor. The number 5, for example, considered either as a written symbol or an abstract concept, is used to capture and represent the mental perception of a certain number of objects—perhaps sheep or onions—considered as an assemblage. From sheep and onions, the number 5 moves to other material assemblages and then to abstract assemblages. Examples of abstract usages: "I can give you five good reasons why Shakespeare's Hamlet is a great play;" "there are only five regular polyhedra" (this does not necessarily imply that we focus on pictures of the five polyhedra).

The mathematical straight line captures, represents, and extends the mental perception of a stretched string. The stretched string of common experience becomes the straight line of Euclid. The Euclidean line leads via Descartes and later writers to a first-degree algebraic equation in two variables, which later leads to the superposition properties of such variables, and then becomes the tremendously broad and important concept of linear analysis and its antithesis, non-linear analysis.

If a simple natural language expression such as "green" has many metaphorical meanings attached to it, in the same way, mathematical expressions such as the straight line or linear analysis can carry many meanings.

2. Mathematical symbolisms as a storehouse of representational possibilities. Mathematical thought, born in verbal expressions, moved gradually to the symbolic, initially as abbreviations for the verbal. The ancient Babylonians may have solved quadratic equations, but it was not until perhaps the 15th or 16th centuries that the algebraic symbolisms that we teach today in high school were developed. The symbolisms themselves are not firm; they have changed over time as the basic ideas that underlie them expand or change. Mathematics has therefore developed its own specialized language of symbols and the rules by which these symbols are allowed to interact with each other and with the non-mathematical world.

The symbolisms, deriving initially from certain concepts, form a supply of standardized expressions that then become part of the mathematician's language. These symbolisms can lose their initial meaning and acquire other meanings; they can go "abstract." Mathematicians adhere to a standard approach to communication that is organized compositionally in certain ways and that then establishes a kind of "social relation-ship" with the targeted mathematical readership. An outsider may feel daunted by the brevity and the sharpness of this sort of discourse that to some extent constitutes a private language.

For these reasons, a given mathematical expression may be applicable to a wide variety of dissimilar situations. The Sapir–Whorf hypothesis, "We make our language and then our language makes us," is relevant here. (We make our automobiles, and then our automobiles shape and change our lives far beyond questions of transportation.) Let me now expand a bit on what I've just said.

As one moves from length to area (the product of two lengths) or to volume (the product of three lengths), one may conceive of the product of four, five, or even more numbers, not now necessarily lengths; if the numbers multiplied are identical, we arrive at the notion of "powers" and their current algebraic symbolisms h, h^2, h^3, h^4, h^5, etc. The second and third powers are called familiarly "squared" and "cubed," revealing their origin in geometry. With such notation at hand, a creative impulse could be to ask "What meaning might be attached to the symbol h^n when n is not an integer, but may be a fraction or a negative number?" The question is answered by bringing in the "addition formula" for suc-cessive multiplications, also called "the law of exponents." By combin-ing of powers through the four processes of arithmetic, one obtains a copious supply of formulas (functions, curves, representations) for the mathematical storehouse or palette.

With these representational possibilities at hand, a scientist or math-ematician, when confronted with a situation from nature, looks among these possibilities for one that mimics, in a certain sense, the natural situation. Thus Galileo's formula $s = gt^2$ represents a stone's fall to the Earth, but it can also represent the growth of the area of the circle that arises when a stone is tossed into a pool of still water.

The algebraic powers and their arithmetic combinations and exten-sions, as just explained, are only one of many, many formulas that mathematics has in its storehouse. The formula that describes exponen-tial growth (the compound interest law) and exponential decay (e.g., of radioactive material) are very well known, so much so that the phrase "exponential growth" has come popularly to mean very rapid and disturbing growth, whether or not the rate of growth is truly exponential in the strict mathematical sense. "Exponential decay" can be heard in publishers' shoptalk as, e.g., when they speak of the "half-life of a new book" on the shelves of bookstores.

Very often a mathematician will try to impose an available formula on exterior events, often with considerable success. This is called "curve fitting," and it is a standard strategy and process of representation and interpretation. The astronomer Johannes Kepler had stunning success with this methodology. The storehouse of mathematical objects includes many different standardized curves, some of which were studied in antiquity. One was the ellipse, and Kepler found that the ellipse was a very good choice to describe the orbit of a planet. On the other hand, F. W. Lancaster was less successful: In 1914, he attempted to describe military conflicts by a square law. It was later found that this doesn't represent the situation very well, and other formulas have been sought.

It is also the case that completely new mathematical languages, expressions, and theories have been developed because of the impact of observational or experimental science. The most familiar example of this is the development of the calculus by Isaac Newton and by Gottfried von Leibnitz to describe variable rates of change.

A non-mathematical example is the modern piano. It has had a slow but constant development from ancient harps in Mesopotamia to today's Steinways or Bösendorfers. A modern piano has 88 keys, and the corresponding strings are tuned in a fixed and widely agreed-upon way. On such a piano we can play hundreds of thousands of different compositions, but there are still musical expressions and harmonies that we cannot obtain. Similarly, the standard supply of mathematical formulas can be invoked in many different situations. Mathematician André Weil summed up the situation by remarking that mathematics provides a palette of objects and operations from which we must pick and choose with care. But new phenomena may call forth new formulas.

3. Mathematics and the unity of nature. One of the grand tendencies of science is to show that nature exhibits a unity of structure and of purpose. Although this idea goes back in history as far as Thales (6th

century BC), in a polytheistic society my feeling is that the current zeal with which it is pursued—at least in the West—derives from religious monotheism.

To exhibit unity means to reduce the number of basic concepts and entities to a bare minimum (in philosophical discussions this principle of action is known as "William of Ockham's Razor") and show that the workings of nature can be explained by a few laws. Comte de Buffon sought the unity of living things and found it in the fact that all forms of life reproduce themselves. Others have proposed an *élan vital* as a unifier, a vital principle, or an animating force within living beings. The astronomer Ptolemy (2nd century) unified the motions of the planets in terms of circles. Mathematicians have sought unity in logic (Bertrand Russell) or in deductive structures as expounded by the French structuralist school of mathematicians known as l'école Bourbaki. Physicists have sought unity in a few basic material objects (elementary particles, etc.) and in a few mathematical formulas that describe their interaction.

Numerous instances tempt one to make claims for unity. Albert Einstein wrote to Marcel Grossman, "It is a wonderful feeling to recognize the unity of a complex of phenomena which to direct observation appear to be quite separate." Consider the physical phenomena that go under the name of the "inverse square law." In algebraic language, $y = k/x^2$ or "y is proportional to $1/x^2$." The inverse square law describes the intensity of light and the forces of gravity, electrostatics, and electromagnetics very well. Is this mere coincidence, or is there an underlying explanation in terms of processes that unifies the four?

When unified processes are represented mathematically, their formulations are similar. Differential equations can act as unifiers. Via Newton's Laws, Pierre Simon Laplace reduced the whole of celestial mechanics to a vast system of ordinary non-linear differential equations. Electrodynamic and hydrodynamic fields are described by one and the same partial differential equation. Some authors have described these equations as "dynamical metaphors."

Contemporary physicists seek an elusive grand unification theory (or GUT) that unifies the strong interaction with the electroweak interaction. Several such theories have been proposed, of which "M-theory" comes the closest. M-theory considers the elementary particles as membranes in an eleven-dimensional space (which would probably make William of Ockham turn in his grave). But no GUT is currently universally accepted. A future theory that will also include gravity is termed a "theory of everything" (or TOE). Once physicists have such a theory, presumably they need theorize no more: What remains to be done is simply to dot a few I's and cross a few T's. In the history of physics, however, this claim has been made before and has not proved out.

Which of these three answers do I personally favor? I prefer "metaphor" and "mathematical storehouse." While I admit that the "unity of nature" is very seductive and often productive, I am not a philosophical monist. In many respects, I am a pluralist, finding pluralism a comfortable philosophy to espouse.

Having said all this, Christina, it occurs to me that my answers to your question "why is it that quite different things can look similar in mathematics?" should be followed by a discussion of what physicist Eugene Wigner has called "the unreasonable effectiveness of mathematics in theoretical physics" and what physicist Arthur Jaffe has called the "unreasonable effectiveness of physics in mathematics." The entire enterprise of applying mathematics to external situations is inherently counterintuitive. Why should the manipulation of a few symbols on a page tell us something about the real world? This comes up not only in the physical sciences but also in social and humanistic applications. To these questions, brilliant people have proposed many different answers. None is adequate and none is universally accepted. But that is for another time and another letter.

Q18. What can an interested amateur do?

A18. Mathematics at the elementary level—addition, subtraction and multiplication—is within the grasp of most people. Some intelligent adults have told me that they have difficulty with fractions (a form of division). When it comes to the higher (or deeper) parts of mathematics—trigonometry, calculus, abstract algebra, for example—although understanding can be augmented by knowledgeable and skillful teachers, comprehension and manipulative skills gradually fall away.

It comes as no surprise that there is a limited number of people who are good at certain things, like figure skating, cartooning, acting, writing, or furniture design, to name a few. These people are professionals in their respective field. It should therefore come as no surprise that there is a limited number of people who can, with professional expertise, comprehend, utilize, and push forward mathematical theories and apply them to problems of physical or social interest Just as a skilled cabinet-maker offers his work for the use and enjoyment of the general populace, the professional mathematician offers his work to the entire public for utilitarian purposes and to a much smaller public for purposes that range from amusement to philosophy to knowledge for its own sake.

Over the years I've received many letters from individuals who are interested in mathematics, who enjoy engaging in it, but who are mathematically unsophisticated. Some of these letters send me attention-getting nuggets that their writers have discovered for themselves. These are usually simple and well-known things, but the writers would like to claim a bit of notoriety for their "discovery." Some letters provide "proofs" or "disproofs" of notorious and well-publicized facts or conjectures. "Circle squarers" are some of these correspondents. (See, e.g., De Morgan's *A Budget of Paradoxes* for the author's encounter with circle squarers.) Others send pages of mathematical-looking symbolisms that, in fact, are nothing of the sort. Professionals are apt to give short shrift to material provided by amateurs.

Occasionally, in his unsophistication (yes, "his"; women correspondents of the above sort are exceedingly rare), a writer will ask a question that comes close to the philosophy of mathematics. I recently received a letter along the following lines:

> Dear Prof. Davis: I have read your book, etc., etc.
>
> I have spent a lifetime at the racetrack, and I have devised the following formula for handicapping horses. [Formula then given, but omitted here.] How can I prove that it's true? Even if it works for a thousand races, does that tell me it's been proved? You don't have to look at a thousand circles of different sizes to prove that their area is πr^2.

This correspondent had devised a formula that seemed commonsensical in view of his track experience. What is interesting, though, is that despite his mathematical naiveté, he raised a question that has been on the forefront of mathematical philosophy for centuries and that, in recent years, comes close to Karl Popper's theory of "falsifiability." The letter intrigued me, and I sent my correspondent a very brief reply:

> Dear Mr. So-and-So:
>
> One must distinguish between the world of theoretical mathematics, which is an ideal, mental world, and the world of real experiences. Within the former world, proof is often possible. Within the latter, experience might make a mathematical formulation plausible, it might disprove it, but it cannot prove it incontrovertibly.
>
> My own racetrack experience is that inside information gained at the paddock, etc., beats mathematical formulas ten to one. Think where the common expression "I have it from the horse's mouth" comes from. Note also that obtaining inside information regarding business matters may be a treacherous path to follow.
>
> Yours, etc.,

Having said all this, I should point out that I have been stimulated by questions raised by professional sculptors who have had little training in mathematics. I have been stimulated by sharp questions raised by schoolchildren. There are books, clubs, and websites for mathematical amateurs. Amateurs can and have raised questions that have led to substantial mathematical developments. Perhaps this has happened more often in the past than in the present, but the possibility exists.

Further Reading

Michel Armatte. "Developments in Statistical Thinking and Their Links with Mathematics." In *Changing Images in Mathematics: From the French Revolution to the New Millennium,* Umberto Bottazzini and Amy Dahan Dalmedico, eds., pp. 137–166. Routledge, 2001.

Julian L. Coolidge. *The Mathematics of Great Amateurs,* 2nd ed. Oxford University Press, 1990. Introductory essay by Julian Gray.

Amy Dahan Dalmedico. "An Image Conflict in Mathematics after 1945." In *Changing Images in Mathematics: From the French Revolution to the New Millennium,* Umberto Bottazzini and Amy Dahan Dalmedico, eds., pp. 223–254. Routledge, 2001.

Philip J. Davis. "What Should the Public Know about Mathematics?" *Daedalus,* Vol. 121, No. 1, Winter 1992.

Philip J. Davis. "Mathematics Is Not Required." *For the Learning of Mathematics*, Vol. 23, March 2003.

Philip J. Davis. "A Letter to Christina of Denmark." *Newsletter of the European Mathematics Society*, March, 2004, pp. 21–24. Also in *Matilde: Nyhedsbrevfor Dansk Matematisk Forening*, Nummer 20, Juni 2004.

Augustus De Morgan. *A Budget of Paradoxes.* Longmans, Green, and Co., 1872; 2nd ed. 1915.

Robert Dewar, e-communication.

Michael O. Finkelstein and Bruce Levin. *Statistics for Lawyers.* Springer-Verlag, 1990.

William James. *Pragmatism: A New Name for Some Old Ways of Thinking.* Longmans, Green, and Co., 1947.

M. MacCrimmon and P. Tillers, eds. *The Dynamics of Judicial Proof, Computation and Common Sense.* Springer-Verlag, 2001.

Jean Mawhin. "Poincaré and Public Affairs." *Notices of the Amer. Math. Soc.*, Oct. 2005, pp. 1041–1042.

Kay L. O'Halloran. *Mathematical Discourse: Language, Symbolisms and Visual Images.* Continuum, 2005.

1 What Is Mathematics?

As an elaboration of Q1–Q5 of "Letters to Christina," it is time to say a few more words about what mathematics is. Strangely, this question is hardly discussed in any course of mathematics routinely taken by undergraduates or graduates. In classes, mathematics is simply what the teacher slaps on the blackboard or displays via PowerPoint.

For starters, we might say that mathematics is the science of quantity and space. This answer might have been satisfactory four hundred years ago, but today we would say that mathematics is multi-faceted: It is the art and science of dealing with deductive (i.e., "theorematic") and algorithmic (i.e., computational) structures that concern themselves with quantity, space, pattern, and arrangement. Mathematics also deals with the language-like symbolisms that allow us to express and manipulate these concepts. It may be noted that these concepts and manipulations have changed over time, often growing, occasionally discarded.

While this definition is more in keeping with the spirit of our times than the brief "quantity and space," some specialists' interest groups—perhaps semioticians, logicians, applied mathematicians, physicists, non-Cantorian set theorists, and others—may feel that what I have just said overemphasizes certain aspects of mathematics at the expense of their own special concerns.

Philosophers of mathematics have also provided definitions of mathematics. Naturally they go for the aspects that are of philosophical interest: epistemology, ontology, theories of knowledge and cognition, semantics, semiotics, etc.

At a higher level of generality, Ruben Hersh and I have stated that "the study of mental objects with reproducible properties is called mathematics." It is a good thing that no definition of mathematics is legislated or comes down by ukase from a mathematical academy or from the dictionary makers. It is a good thing that, unlike doctors, there is no state licensing of mathematicians. This keeps mathematics fluid, as it should be. Mathematics is not in the hands of an academy, nor is it the exclusive property of the community of professional mathematicians. Mathematics lives and is shaped by all who contemplate, speculate on, describe, validate, apply, and develop mathematics.

It is not possible, however, to have mathematics be all things to all people. In any historical age, mathematics is and has to be what the age says it is. The dream of a *mathesis universalis* expressed by René Descartes and Gottfried von Leibniz among others, of an all-embracing mathematical formulation of the representations of the outer physical world, of the decisions, acts, and interpretations of the social world, is unrealistic.

Despite a famous quotation to the effect that "the essence of mathematics lies entirely in its freedom," this should not be interpreted as saying that anything goes in mathematics. Descartes thought he was doing mathematics when he wrote the *Principles of Philosophy,* and for all I know

René Descartes (1596–1650)

Benedict Spinoza may have thought the same about his *Ethica More Geometrico Demonstrata,* which he composed *à la* Euclid with definitions, theorems and corollaries. I don't know whether anyone believed it at the time, but nobody would today.

A hundred advanced texts could not summarize the current state of mathematical knowledge. The *CRC Concise Encyclopedia of Mathematics* runs to almost 2000 pages and merely scratches the surface. A users' handbook for the software package *Mathematica* runs to almost 1500 pages, and computation is only a small part of mathematics.

As in any field, in mathematics there must be both expansion and contraction, or it will become closed and stagnate. Expansion exists in the unfolding of new constructs, new meanings, and new practices. Contraction exists in a variety of ways: as condensation or as the realization or declaration of irrelevance or self-limitation.

By way of illustrating various aspects of mathematics, below are a few examples of statements (or theorems) that display some of the often-overlapping elements of mathematics.

Quantity.

$$13 \times 24 = 312; \; 13 \div 24 = 0.5416666...$$
$$2 - \sqrt{2} = .61254732...$$
$$1 + (1/2^2) + (1/3^2) + (1/4^2) + ... = \pi^2/6$$

Space. A non-circular ellipse has two distinct foci; A and B. The sum of the distances from any point P on the ellipse to the foci is a constant.

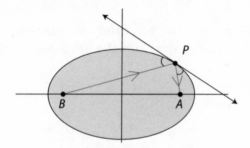

The volume of a sphere of radius r is given by $4\pi r^3$.

A Möbius band is a one-sided surface.

There are only five regular polyhedra: the cube, the tetrahedron, the octahedron, the dodecahedron, and the icosahedron.

Patterns. The elements a, b, and c of a commutative group satisfy the algebraic patterns $(ab)c = a(bc)$ and $ab = ba$.

Pascal's triangle,

$$1$$
$$1 \; 2 \; 1$$
$$1 \; 3 \; 3 \; 1$$
$$1 \; 4 \; 6 \; 4 \; 1$$
$$\text{etc.}$$

is a numerical pattern of integers in which a number in any row is the sum of the two numbers in the previous row and slightly to each side.

The numbers in Pascal's triangle are also known as the binomial coefficients and are found in a great many patterns.

Symmetry. One very important instance of geometric pattern is symmetry. An ellipse has two axes of symmetry.

The five-pointed star below has both rotational and reflective symmetries. If the star is rotated 72°, it will cover the original star exactly. If a line is drawn through the star from one point to the opposite side, and the star reflected over that line, the reflection will look the same as the original star.

The "marigold" figure below is certainly a geometric pattern. It even appears to have rotational symmetry. But this is not strictly the case.

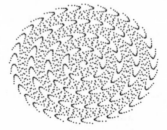

There are logico-algebraic patterns such as

$$\varphi \wedge \psi \equiv \neg\varphi \downarrow \neg\psi$$

which is a symbolic representation of the statement "'both phi and psi' is logically equivalent to the statement 'neither (not phi) nor (not psi)'."

Arrangements. A deck of 52 cards can be arranged in $1 \times 2 \times 3 \times \ldots \times 52$, or 8.0658×10^{67}, ways. Theories of arrangement feed into computations of probability. The probability that a shuffle of 52 cards will produce aces as the top two cards is [4 aces]/[52 cards] × [3 aces]/[51 cards left] = $4/52 \times 3/51 = 1/13 \times 1/17$, or $0.004524\ldots$.

Mathematical structures. A structure is a set of idealized objects that are related to one another in fixed ways or that combine through certain operations. Consider the set N of integers 1, 2, 3,.... Some rules for the combinations of integers are $a + b = b + a$, $ab = ba$, and $a(b + c) = ab + ac$. Within the set N, rules for carrying out the various operations of arithmetic, $+$, $-$, $\times$, $/$, $\sqrt{\ }$, are learned in elementary school and are built into digital computers.

A directed graph is a mathematical structure. What is important in the flow chart below are the nodes and links and not the individual designations given to the nodes.

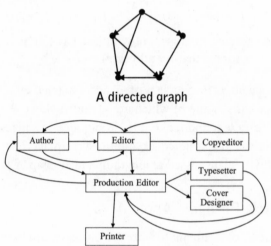

A directed graph

A flow chart as a directed graph

Ruler and compass geometry, wherein points, lines, and circles are allowed to combine in certain specified ways constitutes a mathematical structure.

Mathematical structures carry an aura of exclusivity. They are like social clubs whose laws of admission are set up on the basis of "who shall we keep out?" Thus, the infinite sequence of ones (1, 1, 1,...) is excluded from membership in the Hilbert Space called H^2 because the sum of its squared components is infinite. The purpose of club formation is to insure that members interact properly. Within the club some things are possible for its members and other things impossible.

Deductive structures. The historic and paradigmatic example of this is Euclidean geometry as taught in high school or more sophisticated versions as developed in the late nineteenth century. Here certain "self-evident" statements (axioms) are accepted, from which certain geometric conclusions (theorems) are reached by way of certain accepted rules of logical procedure.

Language-like symbols or notation. Mathematical notations are often found embedded within sentences in natural languages. Here are two clips from technical papers:

$$\text{"}\lim_{n \to \infty} [\varphi(n{+}x) - \varphi(n)] = 0, \text{ for all } x \in (-1, \infty)\text{"}$$

"Let $\psi(v)$ be a forcing arrangement with free variable v such that it's forced that for all $v \; \psi(v) \to v \in \omega$."

Figures. Mathematical discourse is often accompanied by graphical figures of various sorts, some of which are included above. Figures serve not only to clarify text and make it vivid but also to suggest new possibilities. Over the years, however, figures have played an ambiguous role, one that will be discussed at length in a later chapter.

Mathematics: Pure and Applied

Since definitions of mathematics are hardly ever discussed in mathematics classes, it's not surprising that even less discussion is directed toward the distinction between pure and applied mathematics. In elementary mathematical education the two types are intertwined: pure mathematics is manipulations in arithmetic and algebra; applied mathematics consists of "word problems." At higher levels, in college, the situation is somewhat different. A number of universities have two departments of mathematics: pure mathematics and applied mathematics, often at loggerheads.[1] Since the courses offered in these departments occasionally bear the same name, incoming freshmen are often confused as to which course they should take, or, if interested in majoring in mathematics, which type of mathematics they should take.

---

1. Cf. the old German double mathematical joke: there are two kinds of mathematics: *reine* (= *abgewandte*) *Mathematik* und *unreine* (= *angewandte*), i.e., pure and impure.

One could describe the difference between pure and applied mathematics as follows: Pure mathematics is inward looking while applied mathematics is outward looking. That is to say, pure mathematics looks to the study and development of its own theories and interpretations while applied mathematics stresses the application of mathematical theories to the physical and social worlds.

An example from pure mathematics is the theorem of Niels Henrik Abel that the roots of a polynomial of degree five or higher cannot, as a rule, be expressed in terms of integers subjected to the operations of elementary arithmetic and simple root extraction. This theorem (or fact) is of little concern to an applied mathematician.

In contrast, applied mathematics is close to physics, engineering, technology, and economics. Since these subjects are developed via mathematical language, there may be little distinction between them and applied mathematics. Applied mathematics, physics, econometrics, psychometrics, jurimetrics, and technology develop a large variety of mathematical theories or models for dealing with the subjects. If an applied mathematician comes up with a numerical scheme (an algorithm) for modeling the movements of a tornado, it may be of little interest to the pure mathematician. Then again, it may suggest some problems for the pure mathematician to tackle.

Another contrast: Pure mathematics is often pursued with an "art for art's sake" motivation or to escape from the bitter and seemingly unsolvable problems posed by life. Applied mathematics has a utilitarian (in the humanistic sense) motivation. Nonetheless, each type of mathematical concern draws on and contributes to the other, and the boundaries between the two fields are both fuzzy and flexible. The step from the "fantasies" of pure mathematics to real applications in engineering, in the sciences, in social organizations, can be small. Since the pursuit of both the pure and the applied requires financial support, pure mathematics approaches the public for financial support with the reasonable claim that its products are "potentially" applicable. It should not, then, come as a surprise that the relationships between common sense and these two "varieties" of mathematical activity exhibit rather different features and call into play different aspects of common sense.

Further Reading

Amy Dahan Dalmedico. "An Image Conflict in Mathematics after 1945." In *Changing Images in Mathematics: From the French Revolution to the New Millennium,* Umberto Bottazzini and Amy Dahan Dalmedico, eds., pp. 223–254. Routledge, 2001.

Philip J. Davis and Reuben Hersh. *The Mathematical Experience.* Birkhäuser Boston, 1980.

Philip J. Davis and Reuben Hersh. "Rhetoric and Mathematics." In *The Rhetoric of the Human Sciences*, John S. Nelson et al., eds. University of Wisconsin Press, 1987.

J. Hoffman, C. Johnson, and A. Logg. *Dreams of Calculus.* Perspectives on Mathematics Education, Springer, 2004.

Didier Norden. *Les mathématiques pures n'existent pas.* Paris: Actes Sud, 1981.

David Park. *The Grand Contraption: The World as Myth, Number and Chance.* Princeton University Press, 2005.

2 Mathematics and Common Sense: Relations and Contrasts

What Is Common Sense?

Common sense, an expression we all use, is an elusive concept. Is common sense good human understanding or the formulation of evidence based on direct experience? Is it the intuition of the individual? Is it second nature or is it conventional or traditional wisdom? Is it rationality as opposed to irrationality? Is it the opposite of nonsense or of absurdity? Is it all of the above or something more?

Is a person who doesn't use common sense (often called "horse sense") to be considered a fool? Is our brain hard wired for common sense? Is common sense an individual matter or a group matter? Can common sense be inherited? Can it be taught or strengthened? Can it be programmed?

Let's examine the expression.

Language dependence. The elusiveness of the concept is borne out by how the phrase "common sense" translates into different languages. In German, it is *gesunder Menschenverstand,* or sound (healthy) human understanding. Close by, in Denmark, it is *sunde fornuft,* which has about the same meaning as the German. In German-speaking Austria, it is *Hausverstand,* which gives the expression a somewhat different spin. French offers us two different possibilities: *sens commun* and *bon sens.* Hebrew gives us *sechel shachar,* literally, the "sense of the morning." The distinction between common sense as experience and common sense as an abstract idea is implicit in some of these languages, although the distinction is hardly implied by the English phrase.

Place dependence. Every child knows that if he lets a ball go, it will fall to the ground. But what happens to a ball in an orbiting space station? In that environment, "common sense" no longer applies, for the ball may remain in a fixed position.

Culture (i.e., group) dependence. Consider the explanation given to the anthropologist Clifford Geertz by the family of a Javanese boy who fell out of a tree and broke his leg: "The spirit of his deceased grandfather pushed him out because some ritual duty toward the grandfather had been inadvertently overlooked." This is the common sense of the matter.

Time dependence. In antiquity, the passage of the sun and the moon across the sky was observed, so these bodies were thought to go around the earth. At least, this was the inference, and it was common sense at the time to assert it.

Correctability. In time, observations and assertions about the relationship of the Earth to the sun changed. We now know that the Earth goes around the sun.

Gender dependence. Women consider it common sense to ask for directions if they get lost while driving. It is generally acknowledged that, while admitting that it might be sensible to do so, men rarely ask for directions.

Professional dependence. Group goals and experiences enter here. You and I, confronted with a locked door to be opened, would probably not put on a pair of gloves. On the other hand, it would be common sense for a thief to do so. Some statements of quantum theory strike most people as nonsensical, while quantum physicists accept them as part of their every-day thinking.

Semantic dependence. Matthew Westra, quoting a *Wall Street Journal* article: "[A computer manufacturer] is considering changing the command 'Press any key' to 'Press Return Key'. Why? Because many customers have asked 'Where is the Any key?'"

The world is increasingly full of non-verbal signs whose import we must interpret, such as road signs, computer icons, and special squiggles drawn by comic strip artists to indicate motion or emotional states. Many of these can be interpreted on the basis of common sense. Other signs are so arcane, arbitrary, or ambiguous that one really needs to be supplied with a glossary. Not long ago I set off an alarm in the National Gallery in London because I confused an exit sign with a men's room sign.

The distinction between common sense and (mere) information. The distinction between common sense and (mere) information is often vague. In a book to help students achieve quantitative literacy, I found this sample question in a section headed "Numerical Common Sense":

> The average house price in the United States is closest to which number?
> (a) $75,000 (b) $120,000 (c) $185,000 (d) $420,000

Contrary to the implication in the section title, this question requires information, not common sense, to answer it. And note the time dependence; the average house price in the United States is much higher today than it was 100 years ago.

How Philosophers View Common Sense

Philosophers have written much about common sense. Let's tap into a few of their opinions.

Giambattista Vico (1668–1744) introduced a new concept, the *sensus communis* as collective experience and judgment.

> Common sense is judgment without reflection shared by an entire class, an entire nation, or the entire human race.
>
> —Vico, *Scienzia Nuova*, ¶ 142 (quoted by Amos Funkenstein)

The philosopher David Hume (1711–1766) said that common sense equals subjectivity plus the projection onto the other. I suppose he meant by this that common sense is the personal experiences that find echoes or confirmation in the experiences of other people.

The mathematician, physicist, and philosopher Charles Sanders Peirce (1839–1914) explained common sense in this way:

> Common sense notions exist as belief habits that are impervious to doubt when criticized. Example: we believe in the "order of nature." But when-

ever we try to make this belief precise we wind up with disagreement. Yet: who can think that there is no order in nature?

Contemporary mathematician and philosopher Reuben Hersh puts it this way:

> To me, common sense has almost two opposite connotations. The first is col-loquial. If you want to call someone a fool, you say he/she has no common sense. Common sense is knowledge and understanding that doesn't come from schooling, but from simple, practical life. It's common sense to fix the hole in the bucket if you want to use it to carry water. In arguing, people say that their position is just plain common sense. This is a debating trick.
>
> The opposite usage comes in science education, especially in physics, though it can come up in mathematics in a similar way. There, common sense is not a term of praise but almost an insult. Scientific advances have many times violated common sense, have required common sense to be revised, sometimes against heavy resistance. A flat earth used to be com-mon sense. The sun going around the earth, ditto. No need to spell out the details with respect to quantum mechanics and relativity. Here common sense is reduced to little more than unthinking habit. The physicist Wolf-gang Pauli, famous for his "exclusion principle," once made a remark that certain proposed theory couldn't possibly be right because it wasn't crazy enough.

Picking up on Vico, and referring to various views on the philosophy of mathematics, Carlo Cellucci says that the "dominant view" has become so strong that it is identical to common sense.

Pragmatist William James (1842–1910), one of my favorite philoso-phers, points out that our sense impressions have created a number of categories of thought that are inextricably woven into common sense ideas of the way the world is ordered. Among these categories are things, the same or different, kinds, minds, bodies, one time [!], one space, subjects and attributes, causal influences, the fancied, the real.

Of course, there are different takes on these categories. Considering, e.g., "thing," George Berkeley put forward a subjective idealistic view. The common sense view is that the eye perceives things (i.e., objects) whose existence depends in no way on their being observed. On the other hand, quantum physics, whose theories deal with matter and the interrela-tion between matter and the perception of matter, where the act of obser-vation affects what is observed, has decided non-Berkeleyan overtones.

Close, in a way, to James' types of mental organization is the idea of common sense as it appears in contemporary theories of artificial intelligence (AI). One of its theoreticians, Ernest S. Davis, has written

> Almost every type of intelligent task—natural language processing, planning, learning, high-level vision, expert-level reasoning—requires some degree of common sense reasoning to carry it out. The construction of a program with common sense is arguably the central problem of Artificial Intelligence.

In contrast to expert knowledge, which is explicit, common sense reasoning is implicit. The development of common sense computer systems involves making this reasoning explicit using formal logic. The common sense domains that Davis singles out for special attention are "quantity, time, space, physics, intelligent agents, and interpersonal reactions."

Semioticist and literary man Umberto Eco takes a different tack:

> [There is] a twilight zone between common sense and lunacy, [between] truth and error, visionary intelligence and what now seems stupidity, though it was not stupid in its day, and we must therefore reconsider it with great respect.

> — *Language and Lunacy*

Finally, here are the thoughts of philosopher Karl Popper (1902–1994).

> The term "common sense"... is a very vague term, simply because it denotes a vague and changing thing—the often adequate and true and often inadequate and false instincts or opinions of many men.

> Any of our common sense assumptions from which we start can be challenged and criticized at any time; often such an assumption is criticized and rejected (for example, the theory that the world is flat). In such a case, common sense is either modified by the correction or it is transcended and replaced by a theory which may appear to some people for a shorter or a longer period of time as being more or less "crazy." If such a theory needs much training to be understood, it may even fail forever to be absorbed by common sense. Yet even then we can demand that we try to get as close as possible to the ideal: All science and all philosophy are enlightened common sense.

It should be clear, then, that the expression "common sense" is used in a variety of ways and that we have to find our way through this variety.

Compare and Contrast Mathematics and Common Sense

Let's begin by comparing mathematics and common sense.

> There is a logic of mathematics, a logic of the natural sciences, and a logic of ordinary life: Are these three the same... or are they different things that have no connection but the name?
> — R. G. Mayor, *Reason and Common Sense*

Here is an anecdote that I picked up on the Internet: When the famous logician Kurt Gödel studied the Constitution of the United States to become a citizen, he found a logical flaw in it that would allow the existence of a dictatorship. He therefore refused to swear to support the Constitution. Albert Einstein and Oskar Morgenstern were his witnesses, and it took them considerable time and effort to persuade Gödel to go through the naturalization process without confusing the authorities.

A possibly more accurate version of this story can be found in John Dawson's biography of Gödel, but the first story shows that, in most peoples' minds, logic and common sense have little to do with one another. Sane people do things that other people find strange and even incomprehensible. Why do some people skydive? Why do some people accumulate money and property far above the needs for comfortable living? Other people, long before the days of the computer, and at very great labor, compiled concordances to the Bible, to Shakespeare, to James Joyce's *Finnegans Wake*. Still other people collect autographs of movie stars or hundreds of china elephants.

Mathematicians very often do strange, seemingly idiotic, things, like, for example, proving what is perfectly obvious to the rest of the world. In the 6th century BC, Thales of Miletus (c. 620–540 BC) "proved" that the diameter of a circle divided the circle into two equal parts. Ridiculous? Who would deny it?

Yet Thales started an intellectual trend that is one of the finest, one of the most stunning, products of human thought and whose end is not yet in sight. (After all, art is often a mixture of the sublime and the ridiculous.) And from that point on, the Ancient Greeks stressed deductive reasoning as a method of mathematical proof. Here are four of the theorems ascribed to Thales:

(1) When two straight lines intersect, the opposite angles are equal.
(2) The angles at the base of an isosceles triangle are equal.
(3) A diameter of a circle divides the circle into two equal parts.
(4) An angle inscribed in a semicircle is a right angle.

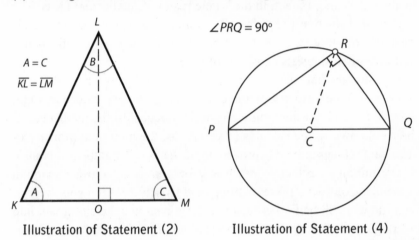

Illustration of Statement (2) Illustration of Statement (4)

Statements (1), (2), and (3) seem so obvious by simply looking at the diagrams that it would be a denial of common sense to think otherwise. Statement (4) is not so immediately obvious. In Cases 1–3, what on Earth is the point of a deductive "proof"? Does it strengthen our belief in the truth of the particular situation? Mathematician and philosopher Alfred North Whitehead (1861–1947) wrote: "It requires a very unusual mind to undertake the analysis of the obvious." Aristotle remarked that to Thales the important question is not *what* we know but *how* we know it.

I will inject a personal story here. Some years ago, I had a correspondence with the famous positivist British philosopher Alfred Ayer. I wrote Ayer a letter and asked him what he now considered to be the most important philosophical questions. He wrote back immediately suggesting three, of which one question was *"What is the case and how do we know it?"* I believe the question of how we know something has become increasingly important in a day when computer simulations and virtual reality increase the level of ambiguity in what is put forward as evidence. The virtual is often confused with common sense.

The Greek followers of Thales went on to elaborate the consequences of the "hypothetico-deductive method" of mathematics to the point where it has produced statements that for the most part are far beyond common sense obviousness. This method was a unique contribution to mathematics and, in fact, to the whole realm of intellectual ideas. The notion of deductive proof from "commonly accepted" axioms lends method, objectivity, certainty, accreditation, and beauty to mathematics and very often suggests new lines of inquiry.

Is a straight line the shortest distance between two points? Common sense appears to say so; there is visual or physical evidence. Light travels in a straight line: more physical evidence. In the mathematical subject known as the calculus of variations, the minimum property of the straight line segment is proved on the basis of a number of axioms. Every minimum principle can easily be converted into a maximum principle. Consider a piece of string held loosely between your fingers; your fingers will be farthest apart when you stretch the string taut into a straight line. But "howevers" or qualifiers rear their disturbing heads and contradict the straight line answer. On the surface of the earth, the shortest distance between two points is not a straight line but an arc of a great circle. On surfaces of different shapes the minimal distances vary. These minimal distance curves are the so-called geodesics, the study of which comprises a long and deep chapter of differential geometry.

Consider, also, surfaces with obstacles. If, when you are out for a walk and want to get to the opposite side of a pond, it would not be common sense to try to do it by walking on water. Mathematics tells us that we might want to build a bridge to get across minimally; common sense requires something else. Consider the shortest walking distances in Manhattan; surely they are not always in a mathematical straight line. Or consider the strange metric topology of airline routings. A routing from Providence to Los Angeles might take me through Atlanta—counterintuitive? On the other hand, the possibility of certain connections might tell me that it would be common sense for me to backtrack from Providence to Boston and fly from there. In a complex of buildings, architects and landscapers will often create formal paths connecting them, but human nature causes people to take short cuts, wearing away grass instead of following the paths.

Now, what about light? The path it takes depends on the nature of the medium through which it passes. Water changes the velocity of light and bends its path (refraction). The minimum principle preserved in this case is the minimum time of passage. Relativity theory (confirmed by experience) says that light can be bent by gravitational fields.

The upshot of these simple observations is that the statement about a straight line being the shortest distance between two points is true only when properly qualified. It might take a very long list to specify these qualifications, and the list would certainly include clarifications of what is meant by the words "straight" and "shortest distance." Furthermore, experience tells us that a common sense view of a situation often leads to human actions that tend to preserve the common sense view.

In the hands of the Greeks, then, mathematics was raised via logical deduction from Babylonian bookkeeping to the status of high art. Russell and Whitehead, after hundreds of pages of their *Principia Mathematica* (c. 1910), proved that $1 + 1 = 2$. Again, high art, as in classical ballet, is often a mixture of the sublime and the ridiculous.

Now for some contrasts.

Mathematical thinking questions its own premises. It constantly reviews the axioms or hypotheses on which it is built. Common sense rarely examines itself.

Common sense is part of the grammar of social intercourse. It is unarticulated and not formalized or given abstract structure, contrary to mathematics. Its lack of articulation notwithstanding, a number of computer scientists have thought that vast numbers of individual instances of common sense can be built into computer instructions. To some extent it can. But when an automatic answering system fails to deal with our particular problem, we all shriek "let me talk to a flesh and blood person!"

(Pure) mathematics tends to free itself of outside contexts. Common sense does not. There can, however, be common sense even when one works in pure mathematics.

Common sense governs the process of abstraction from a real-world situation to a mathematical model. This includes an understanding of

(a) what kinds of calculation of the model are likely to be valid in the real world,

(b) what aspects of the real world are being abstracted away,

(c) to what extent are the calculations uncertain or approximate as carried out, and

(d) how the model can be extended or refined to improve the certainty or precision.

Common sense identifies clearly incorrect conclusions that follow from inappropriate mathematical models, technical errors, or simple mistakes.

The more that mathematics is built into daily practices via "chipification," the less explicit is the mathematical knowledge required on the part of the public. Since chipification is often hard to identify, modify, or delete, ignoring common sense can leave the mathematizations of social life uncontrolled by assessment and feedback. There are innumerable instances of this; war gaming, IQs, district gerrymanders via political software come to mind easily. More recently, *Business Week* of January 23, 2006, under the heading of "Math Will Rock Your World," points out that the mining of personal data of all sorts for a variety of reasons has become a major industry requiring advanced mathematical skills. The bright side of this tendency is better security, better health, and greater profits. One sanguine prediction is that "the next Jonas Salk will be a mathematician and not a doctor." The dark side is the loss of privacy and "a sense that individuals are powerless, a foreboding that mathematics, from our credit rating to our genomic map, spells out our destiny."

If mathematics were simply common sense, it and its applications would be trivial:

> Mathematical biology, to be useful and interesting, must be relevant biologically. The best models show how a process works and predicts what may follow. If these are not already obvious to biologists and the predictions turn out to be right, then you will have the biologists' attention.

—James D. Murray, Oxford biologist

Common sense displays modesty:

> Being a man of the liveliest and most ingenious intellect, [philosopher Isaiah Berlin] must wish that reason could do more to transform the human condition radically and quickly; being a man of common sense he knows it cannot.

—Edward Crankshaw, *The Observer* (British newspaper)

Mathematics, on the other hand, can be "imperialistic." Pythagoras (550 BC) said that "All is number." Current neo-Pythagoreans or mathocrats believe that all problems can be solved by finding an appropriate mathematical formulation or model. The digital computer, a mathematical and informational engine, has "conquered" the world and set up its own imperialistic goals.

Further Reading

Carlo Cellucci. *Filosofia e matematica*. Laterza, Bari, 2002. Introduction available in English.

Ernest S. Davis. *Representations of Commonsense Knowledge*. Morgan Kaufmann, 1990.

Philip J. Davis. "Mathematics and Common Sense: What Is Their Relationship?" *SIAM News*, November 1995.

Philip J. Davis. "Mathematics and Common Sense: Cooperation or Conflict." In *Proceedings of the 47th CIEAEM Meeting: Mathematics (Education) and Common Sense*, Christine Keitel, ed. Freie Universität Berlin, 1996.

John W. Dawson, Jr. *Logical Dilemmas: The Life and Work of Kurt Gödel*. A K Peters, 1997.

Amos Funkenstein. *Theology and the Scientific Imagination*. Princeton University Press, 1986.

Clifford Geertz. *Local Knowledge: Further Essays in Interpretive Anthropology*. Basic Books, 1983.

Reuben Hersh, personal communication.

John Horgan. "In Defense of Common Sense." *NY Times*, Op-Ed, Aug. 12, 2005.

William James. "Pragmatism and Common Sense." In *Pragmatism: A New Name for Some Old Ways of Thinking*. Longmans, Green, and Co., 1907.

Christine Keitel. "Mathematics Education and Common Sense." In *Proceedings of the 47th CIEAEM Meeting: Mathematics (Education) and Common Sense*, Christine Keitel, ed. Freie Universität Berlin, 1996.

Niels Møller Nielsen. "Counter Argument: In Defense of Common Sense." Thesis, Department of Language and Culture, Roskilde University, Denmark, 2000.

Karl Popper. *Objective Knowledge*. Oxford University Press, 1972.

Eric W. Weisstein. *CRC Concise Encyclopedia of Mathematics*. Chapman & Hall/CRC, 1998.

How Common Sense Impacts Mathematical Reasoning: Some Very Simple Problems

When mathematical reasoning is used to solve a real-world problem or to gain information about a real-world situation, an interaction takes place between the formal mathematical operations and an underlying commonsensical understanding of the world. This interaction can be very complex and has many different aspects:

- Common-sense understanding helps us decide whether a real-world problem can be translated into a mathematical question about an abstract model.
- Common-sense understanding helps us choose the appropriate mathematical model for a problem; it helps us decide which features of a real-world situation should be abstracted and reflected in the model and which may be ignored, and helps us determine to what extent calculations are uncertain and approximate.
- Common-sense understanding helps us interpret the answers given by mathematical calculations.
- When mathematical calculations yield an answer that seems contrary to common sense, a complex process of evaluation takes place. The conflict may be resolved in many different ways: a technical error in the calculations may be found; it may be decided that the mathematical model does not apply to the situation or does not give correct results for the problem; alternatively, it may be decided that the mathematics is right and the common sense intuition is wrong, and this may lead to further refinements and improvements of the reasoner's common-sense knowledge and intuition.

Consider an old computational chestnut: A person has two parents, four grandparents, eight great-grandparents, etc. After 50 generations, which is perhaps 1500 years, we might infer that there were quadrillions of people in the world. This is very unlikely; we eventually conclude that, over the generations, there have been some marriages of cousins, more marriages of second cousins, and so forth.

Below are thirteen simple mathematical "models" and their common sense underlay provided to me by Professor Ernest S. Davis. How a real-world problem is translated into mathematics depends on the situation and on the problem to be solved; the relation is complex and often subtle. The examples that follow show how a generic geometric problem—"How many X's fit into Y?"—can require many different forms of mathematical analysis.

Example 1. There are fifty-five chairs in a classroom. Thirty of these are currently occupied. How many more people will fit?

Mathematical solution: $55 - 30 = 25$

Discussion: The analysis in the example assumes the chairs are slots to be counted. Note the implicit assumption that each person occupies exactly one chair.

Example 2. How many books, each 1 1/4 inch thick, will fit on a 36-inch-long shelf?

Mathematical solution: $\left\lfloor \dfrac{36}{\frac{5}{4}} \right\rfloor = 28.$

Notation: $\lfloor x \rfloor$, called the "floor" function, designates the greatest integer less than or equal to the number x.

Discussion: Since we don't usually count half-books (for example), we round down to the nearest integer, or whole book, rather than use the exact answer $36/(5/4) = 28.8$. The analysis assumes that the spaces occupied by two books do not overlap and that a book maintains a constant shape but can be moved and rotated.

Example 3. How many glasses of water, each 1 1/4 cups, can be poured into a one-quart pitcher?

Mathematical solution: $\dfrac{4 \text{ cups per quart}}{\frac{5}{4} \text{ cups per glass}} = 3.2$ glasses.

Discussion: In this case, it makes sense to speak of parts of a whole. The analysis would be the same if the question were rephrased to

"How many glasses, each 1 1/4 cups, can be poured out of a one-quart pitcher?" A glassful of water does not maintain its shape or even its separate identity when it is combined with other glasses in a pitcher; nonetheless, the volume occupied by the combined glasses is the sum of their individual volumes.

Example 4. How many glasses of water, each 500 cubic centimeters, can be poured into a pot whose interior is a right circular cylinder with a radius of 10 cm and a height of 40 cm?

Mathematical solution: The volume of the interior of the pot is $\pi \times 10^2 \times 40 = 12{,}566.4$ cc. It will hold $12{,}566.4/500 = 25.13$ glasses.

Discussion: This example is similar to Example 3, except that the volume of the container must be calculated from its dimensions. The formula for volume of a right cylinder is taught as a "cookbook" formula in high-school geometry, although deriving the formula requires some form of the integral calculus. In evaluating the significance of this answer, the reasoner should realize that, unless these measurements and pourings are carried out with extraordinary care, the "13" after the decimal point is meaningless. If we were to perform this experiment, it is quite likely that the pot would overflow at the 25th cup, or that a fraction of the 26th cup significantly more than 0.13 could be added at the end. The answer should read "about 25 cups."

Example 5. If a camel can carry up to 500 pounds, how many one-and-a-half-pound books can be loaded onto it?

Mathematical solution: $\dfrac{500 \text{ pounds}}{1\frac{1}{2} \text{ pounds per book}} = 333$ books.

Discussion: The calculation is similar to the calculation in Example 2; the weight the camel can carry is divided by the weight per book, and the result is rounded down because we are only interested in whole books. Unlike Example 2, however, there is no geometric underpinning here; the calculation is based on the natural law that the weight of a collection of objects is the sum of their individual weights. The common sense reasoner will realize that how much a camel can carry depends on the circumstances surrounding the situation: How far? How fast? Over what kind of terrain? Etc. The statement that the camel can carry up to 500 pounds is a nominal value attached to an incompletely defined quantity. Also, the calculation ignores the weight of the boxes and ropes needed to load the books onto the camel. The answer therefore should read "Probably about 300 books, or somewhat more."

Example 6. A monthly calendar is laid out in a rect-angular array, with 1 column for each day of the week from Sunday to Saturday (see diagram). How many rows are needed to fit an *N*-day month?

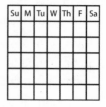

Mathematical solution: Either $\lceil N/7 \rceil$ or $\lceil (N+6)/7 \rceil$, depending on what day of the week is the first of the month.

Notation: $\lceil x \rceil$, the "ceiling" function, designates the smallest integer greater than or equal to the number *x*.

Discussion: Most people will require a little experimentation to find this formula.

Example 7. A bookcase contains 5 shelves, each 36 inches long. How many books, each 1 1/4 inches thick, can be fit in the bookcase?

Mathematical solution: Using the results of Example 2,

$$5 \times \left\lfloor \frac{36}{\frac{5}{4}} \right\rfloor = 140.$$

Discussion: Note that it would be incorrect to do the following com-putation: There are 5 × 36 = 180 inches of shelf space; hence I can fit 180/(5/4) = 144 books. The reason, of course, is that a book cannot be split between two shelves. (It is precisely that "of course" and the sense in which this observation is a reason for the form of the calculation, that is the meeting place between common sense and mathematics.) The cal-culation here, however, does give a useful approximation to the correct answer, and a reliable upper bound for the correct answer. Both these observations are further aspects of the common-sense understanding of mathematical modeling.

Example 8. A bookshelf is 36 inches long, 10 inches high, and 6 inches deep. I have a collection of books, each 1 1/4 inch thick, 8 inches high, and 4 1/2 inches wide. How many of my books will fit on the shelf, if the books may be placed vertically, next to each other, or horizontally, stacked one on top of the other, with the spines visible from the front?

Mathematical solution: You can make four horizontal stacks of books, each 8 books high, and then place 3 books vertically at the end, for a total of 35 books.

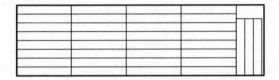

Discussion: The situation is now conceptualized in terms of a two-dimensional layout. (Because of the constraint that the spines must be visible, the four-and-a-half-inch width of the books is irrelevant, as long as it is less than the depth of the shelf.) The correctness of this solution can be checked by simple arithmetic. Finding the optimal solution for this kind of problem can be difficult.

Example 9. Revise Example 8 to drop the requirement that the books' spines be visible from the front. (This, alas, corresponds to the situation in the author's own bookcase.)

Mathematical solution: Place the first 35 books as in Example 6, then place 8 books flat behind the stacks (at the back of the shelf), for a total of 43 books. (See diagram.)

Stack of 8	Stack of 8	Stack of 8	Stack of 8	

Discussion: The situation is now conceptualized in terms of a three-dimensional layout. Like Example 8, this solution is easily checked using arithmetic.

Example 10. Will a 5-foot diameter circular tablecloth cover a 3-foot-square table?

Mathematical solution: The four corners of the table are the farthest points from its center and the distance from a corner of the table to the center is $\frac{3\sqrt{2}}{2} = 2.1$ feet. The radius of the cloth is $5/2 = 2.5$ feet. Therefore, if the center of the cloth is placed over the center of the table, the cloth will cover the table.

Discussion: As in Example 8, the problem is conceptualized in terms of two-dimensional geometry; it requires high-school geometry to solve.

Example 11. Can three 1-foot diameter soccer balls be placed inside a 1 foot 10 inch cubical box?

Mathematical solution: Place the three spheres at the space coordinates $(0, 0, 1/\sqrt{2})$, $(0, 1/\sqrt{2}, 0)$, and $(1/\sqrt{2}, 0, 0)$. Using the Pythagorean theorem, the distance between the centers of any two balls is 1.0; since each ball has a radius of 0.5, this placement does not cause any two balls to overlap. The least coordinate in the boundary of any of the balls in any dimension is –0.5. The greatest coordinate is $0.5 + 1/\sqrt{2}$. Hence, a cube with sides parallel to the coordinate axes and side length $1 + 1/\sqrt{2} = 1.707$ can be drawn around the balls. A cube with side length 1 foot 10 inch = 1.83 feet is definitely large enough to hold the balls.

Discussion: This is a problem in three-dimensional geometry within the scope of high-school geometry. Finding the maximal number of spheres, or other shapes, that will fit inside a given container is known as the "packing problem." It has been much studied by mathematicians, but many basic questions about it have not yet been resolved theoretically.

Example 12. Show that a 5-foot-square tablecloth can be fit inside a 2 foot 6 inch by 1 foot 3 inch rectangular drawer folding the cloth no more than 3 times.

Mathematical solution: Fold it twice vertically and once horizontally.

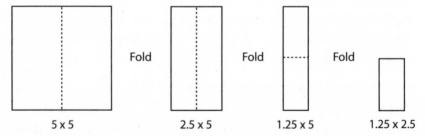

| 5 x 5 | 2.5 x 5 | 1.25 x 5 | 1.25 x 2.5 |

Discussion: For a reasoner who understands how folding works, checking this solution is a trivial mathematical exercise, requiring two simple divisions. What is not trivial is to construct a mathematical model that relates the number 3 (as in "three folds") to this. To do so, you need the concept of an operator that transforms one shape to another and you need to define the result of "folding" shape S at line L.

Definition: Let S be a two-dimensional shape and let L be a line through S. Let S_{left} be the part of S on one side of L and let S_{right} be the part of S on the other side. Then the result of folding S at L is the union of S_{left} with the reflection of S_{right} across L. The problem can then be stated as follows: Let S_0 be a 5-foot square. Find lines L_1, L_2, and L_3 such that fold(fold(fold(S_0, L_1)L_2)L_3) fits inside a 2 foot 6 inch-by-1 foot 3 inch rectangle.

The physical constraints on foldable objects are quite different from the constraints on rigid objects like books. (Strictly speaking, of course, a book is not a rigid object, and if you are willing to contemplate solutions to the problem of fitting books on shelves that involve having them placed open on the shelves, then in some cases you can fit in a few more books.)

Example 13. Show how a 2-foot cube can be cut into four pieces that can be rearranged to fit inside an $8 \times 1 \times 1$ box.

Mathematical solution: Cut the cube into four $2 \times 1 \times 1$ boxes (dimensions in feet), as shown.

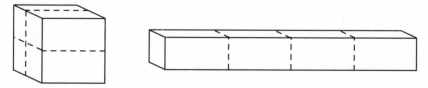

Discussion: Like Example 12, this requires an understanding of the nature of a process that transforms the shapes of objects; in this case, the cutting operation.

The moral of this chapter is that the common sense information required to arrive at mathematical solutions to these "simple" problems is considerable. Much more would be required when the problems to be modeled are not so elementary.

Reference

Ernest S. Davis, e-communication.

Where Is Mathematical Knowledge Lodged, and Where Does It Come From?

The questions "where is mathematical knowledge lodged?" and "where does it come from?" are independent, and no time sequencing should be inferred. According to how we answer them, we will affect how we conceptualize, how we mathematize, how we do research, and how we teach.

In the pre-Cartesian age, knowledge was often thought to be lodged in the mind of God. It was asserted that knowledge derived from authority, reason, and experience, but experience was downplayed. Those who imparted knowledge authoritatively derived their authority from their closeness to the Mind of God, and this closeness was indicated by their personal godliness. In the post-Cartesian age, knowledge was thought to be lodged in places that are above and beyond ourselves, such as sound reasoning or creative genius or in "the object objectively known." A recent view, connected perhaps with the names of Thomas Kuhn and Imre Lakatos, is that knowledge is socially justified belief. In a parallel assertion, R. H. Brown has asserted that reason itself is socially constructed.

Today, a mathematician is more likely to answer the question pragmatically by saying that mathematical knowledge is lodged in tables, books, periodicals, documents, databases, "chipified" computer programs and automata, and software libraries, as well as in the individual skills of mathematicians, their memories, experimental results, and group interactions.

The backlog of known mathematics is vast. The current production of new mathematical results, theories, and techniques is beyond any individual's ability to absorb. Common sense suggests that making this trove of knowledge accessible, arranged and pinpointed in a usable fashion, malleable toward the production of applications and of additional mathematics, would be an enterprise of the utmost value. This, of course is predicated on the assumption that civilization recognizes in the mathematical spirit a natural expression of human creativity as well as a road to further human progress and a way of relieving human dilemmas.

A recent subfield of artificial intelligence (AI) is dedicated to the problem of mathematical knowledge management (MKM). The problem itself is ancient. There are mathematical tables that were made 4000 years ago. The problem of discovering what is known, if anything, about a concept and how it connects with other material is not an easy one, even when the scope of the management solution has been severely restricted. One of the most ambitious MKM projects is the *NIST Digital Library of Mathematical Functions* (NIST stands for the National Institutes of Standards and Technology). The project deals only with special mathematical functions, such as the exponential function and the Bessel functions, that are essential in engineering, physics, and statistics. In the words of Daniel Lozier, the principal investigator,

> The mathematical content is being provided by expert authors... and its website will have facilities to search for text and equations, extract formulas and other mathematical data reliably, display user-controllable visualizations of functions, obtain reviews and full texts of references electronically, and link to sources of relevant mathematical software.

There are numerous difficulties on the road to automating MKM. Among the many problems that interfere (thankfully) is the ultimate necessity of assessing the scientific or human worth of what has been created.

There is hardly a culture that has not developed some kind of mathematics. In the west, newly developed mathematics is produced by "lone riders," by people in university departments, in scientific and technical and organizations. Mathematics arises from the pressures and potentialities of the government, of the business world, of the military, of the artistic world, of the entertainment world, and of day-to-day living.

The answer to the question of where mathematics comes from depends, of course, on what one means by "comes from." I have heard all kinds of answers given to this question with the possible exception of "the stork brings it." Here are a few answers that I have heard: mathematics is nothing but common sense; mathematics comes from God; mathematics comes from the fact that the universe "speaks" mathematically. I have also heard that it comes from people's experience with the natural world and from their ability to abstract imaginatively, to generalize, and to make and manipulate symbolic representations of what they have experienced; it comes from people's ability to create metaphors; it comes from the fact that people's brains are "hard wired" in certain ways for mathematics.

The average person does not really think about the nature of mathematics. He or she just carries out mathematical operations as the need arises and as ability permits. Professional mathematicians, on the other hand, do think about the matter and often come up with conflicting views.

George Lakoff and Rafael Núñez, distinguished cognitive scientists, say

In recent years, there have been revolutionary advances in cognitive science—advances that have an important bearing on our understanding of mathematics. Perhaps the most profound of these insights are the following:

(1) The embodiment of the mind. The detailed nature of our bodies, our brains, and our everyday functioning in the world structures, human concepts and human reason. This includes mathematical concepts and mathematical reason.

(2) The cognitive unconscious. Most thought is unconscious—not repressed in the Freudian sense but simply inaccessible to direct conscious introspection. We cannot look directly at our conceptual systems and at our low-level thought processes. This includes most mathematical thought.

(3) Metaphorical thought. For the most part, human beings conceptualize abstract concepts in concrete terms, using ideas and modes of reasoning grounded in the sensory-motor system. The mechanism by which the abstract is comprehended in terms of the concrete is called conceptual metaphor. Mathematical thought also makes use of conceptual metaphor, as when we conceptualize numbers as points on a line.

the abstract is comprehended in terms of the concrete is called conceptual metaphor. Mathematical thought also makes use of conceptual metaphor, as when we conceptualize numbers as points on a line.

Lakoff and Núñez give a list of the cognitive capacities required for doing simple arithmetic. They are (with these authors' elaborations omitted):

> ... grouping capacity, ordering capacity, pairing capacity, memory capacity, exhaustion detection capacity, cardinal number assignment, independent-order capacity, combinatorial-grouping capacity, symbolizing capacity, metaphorizing capacity, conceptual-blending capacity.

Reading this list, I begin to feel like Molière's Monsieur Jordain who didn't realize he was speaking in prose. I was never fully conscious that in doing arithmetic I was invoking all these capacities. Then again, I realize that there may be some people who lack some of these capacities. For example, I have often wished that in my professional work my memory had been better. But there is one very important capacity that I would like to add as an amplification of this list: the capacity to recognize when *not* to use the various operations of arithmetic. This knowledge is based on experience and then becomes part of the common sense of arithmetic. I develop the necessity of this last capacity more fully in Chapter 7, "Why Counting Is Impossible," and Chapter 9, "When Should One Add Two Numbers?"

Lakoff and Núñez also put forward the notion of "grounding metaphors" as a source of mathematical ideas. The metaphors they single out are attractive and didactically useful, but they are also problematic and have accordingly attracted flak. They say that simple arithmetic is "grounded" in the experiences of the collection of objects, combining objects together, stringing segments together, and motion on a line, but they sometimes misconstrue the structure of the metaphor. Thus, in the "measuring stick metaphor," they associate "the basic physical segment" with the number 1. But of course "the basic physical segment" is just a back-construction from the concept of "one." Another example: The relation $A = C \div B$ is metaphorized by "the repeated subtraction of physical segments of length B from an initial physical segment of length C until nothing is left of the initial physical segment. The result, A, is the number of times the subtraction occurs." But this operation is not one that naturally arises in dealing with measuring sticks.

> In their account of propositional logic, Lakoff and Núñez put too much emphasis on the imperfect analogy with physical containers and much too little on the relationship with predication. When you hear that "Sam is a guppy" and infer that "Sam is a fish"—using the known fact that all guppies are fish—and infer that "Sam is not a dog"—using the fact that no fish are dogs—you are not visualizing little jars marked "guppies", "fish" and "dogs". You are reasoning directly about the properties. The structure of propositional logic comes mainly from examples like these, and only secondarily from visualizing containers or Venn diagrams.
>
> —Ernest S. Davis, in *Mathematics as Metaphor*

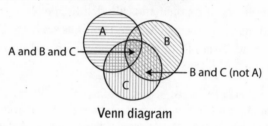

A and B and C

B and C (not A)

Venn diagram

In a paragraph entitled *What Is Special About Mathematics?* Lakoff and Núñez write

> As subsystems of the human conceptual system, arithmetic in particular and mathematics in general are special in several ways. They are: precise, consistent, stable across time and communities, understandable across cultures, symbolizable, calculable, generalizable, and effective as general tools for description, explanation, and prediction in a vast number of everyday activities from business to building to sports to science and technology.

While agreeing that mathematics is a special system of concepts, I would express only partial agreement with the above sentiments—even for pure mathematics. They express dreams of an ideal discipline. Let me critique several of these adjectives.

Precise. We humans cannot even count large groups of objects with precision (See Chapter 7, "Why Counting Is Impossible.") And note

Stable across time. Yes, more or less; but errors, limitations, and exceptions are found. Errors are worked on, and basic assertions are changed as a consequence.

Understandable across cultures. Yes, when we limit the discussion to current "normative" mathematics. Old Polynesian mathematics may not be totally translatable into our terms.

Specialists' mathematical theories may not be understandable across the hallways of university departments of mathematics. The mathematics of the theories of turbulence of fluids may not be understandable to those whose research concentrates on recent advances in symbolic logic. One might claim that "in principle" the two cultures can cross-communicate, but in practice this would be difficult.

The cognitive views of mathematics alluded to above have one virtue: The mind with its various capacities and abilities seems to be one mind. On the other hand, Brian Rotman, mathematician and semioticist, posits a mathematical mind that is a trinity:

> The mathematician has three bodies, or three material arenas of operation—a mortal person, a virtual agent, and a semiotic subject—likewise co-assembled. The mathematical *person* subjectively situated in language is the one who imagines, makes judgments, tells stories, has intuitions, hunches and motives; next, the mathematical *agent*, imagined by the person, is a formal construct which executes ideal actions and lacks any capacity to attach meaning to the signs which control its narratives; and between them, their interface, the mathematical *subject*, who embodies the materiality of the apparatus that writes and is written by mathematical thought.

It is beyond the scope of this essay to show how these three personas cooperate in specific instances such as dealing with bookkeeping operations, various infinities, or the subtleties of mathematical logic. From my own experience, I can assert that they are occasionally at daggers drawn.

While giving pride of place to internal forces and to physical phenomena, histories of mathematics often give short shrift to the historical and cultural elements at work in the development of the subject. Yes, mathematics may be just common sense, or it may be divinely inspired, or the physical universe may "speak" in mathematical terms, or it may be a vast and remarkable metaphor, but mathematics also lies embedded in human history and culture. The accidents of time, place, and potentiality have forced certain characteristic features onto the subject, but to assert historic inevitability for a mathematical concept is problematic.

The impetus to mathematize can come from trade, business, the stock market, warfare, insurance, human biology, chemistry, psychology, social complexities, law, technology, and the physical universe from the micro to the macro level, which have all shaped and continue to shape the course of mathematics. Reciprocally, mathematics, in combination with these, forces changes in human behavior and arrangements. Artificial satellites, for example, dependent upon Newton's laws and an evolving calculus and a highly developed mathematics of dynamical systems, combine with Maxwell's laws of electromagnetism, result in TV, military surveillance, the Internet, and cell phones, all of which have inundated the world in tsunami-like fashion, affecting human behavior in substantial ways. There is nothing inevitable to people talking in the street, in supermarkets, to their counterparts who may be thousands of miles away in their beds. It was not done before. Part of the interaction between mathematics and the collective common sense—if there is such a thing—is to know when to say "yes" and when to say "no" to the seductive products that mathematics helps to bring forth.

Further Reading

Marcia Ascher. *Mathematics Elsewhere: An Exploration of Ideas across Cultures.* Princeton University Press, 2002.

Bernhelm Booss-Bavnbek and Jens Høyrup, eds. *Mathematics and War.* Birkhäuser, 2003.

Kenneth Bruffee. "Social Construction, Language, and the Authority of Knowledge: A Bibliographical Essay." *College English*, Vol. 48, 1986.

Ernest S. Davis. "Mathematics as Metaphor: Review of 'Where Mathematics Comes From.'" To appear in *The Journal of Experimental and Theoretical Artificial Intelligence.*

Stanislas Dehaene. *The Number Sense: How the Mind Creates Mathematics.* Oxford University Press, 1997.

Jean-Pierre Dupuy. *The Mechanization of the Mind: On the Origins of Cognitive Science.* Princeton University Press, 2000.

Paul Ernest. "Forms of Knowledge in Mathematics and Mathematics Education: Philosophical and Rhetorical Perspectives." *Educational Studies in Mathematics,* Vol. 38, 1999.

Torkel Franzén. *Gödel's Theorem: An Incomplete Guide to Its Use and Abuse.* A K Peters, 2005.

David Henderson. "Review of George Lakoff and Rafael Núñez." *Mathematical Intelligencer,* Vol. 24, Winter 2002.

Reuben Hersh. *Eighteen Unconventional Essays on the Nature of Mathematics.* Springer, 2006.

Thomas S. Kuhn. *The Structure of Scientific Revolutions,* 2nd ed. University of Chicago Press, 1970.

Imre Lakatos. *Proofs and Refutations.* Cambridge University Press, 1976.

George Lakoff and Rafael Núñez. *Where Does Mathematics Come From: How the Embodied Mind Brings Mathematics into Being.* Basic Books, 2000.

Brian Rotman. *Mathematics as Sign: Writing, Imagining, Counting.* Stanford University Press, 2000.

Brian Rotman. *Ghost Effects.* Address presented at Stanford Humanities Institute, November 2004.

Raymond L. Wilder. *The Evolution of Mathematical Concepts.* John Wiley & Sons, 1968.

Leslie White. "The Locus of Mathematical Reality: An Anthropological Footnote." *Philosophy of Science,* October 1947.

5 What Is Mathematical Intuition?

Wenn ihr's nicht fühlt, ihr werdet's nicht erjagen.
(If you don't feel it, you won't be able to grasp it.)
—Johann Wolfgang von Goethe

The only thing about mathematical intuition that can be stated with certainty is that it is important.
—Saul Abarbanel

We have a primitive intuition about numbers, space, etc.
—Kant, and many others

My gut feeling, based on what I know, is that there are an unlimited number of twin primes: 3 and 5, 5 and 7, 11 and 13...
— from a number theorist working on the problem

My intuitive feelings about the way that the mathematics ought to work come from my knowledge about the way things actually work.
—from a mathematician/engineer

There is a great philosophical, metaphysical, cognitive, physiological industry devoted to discussing the nature of intuition. Is intuition the experience of the "aha!" or a dramatic illumination on the road to a mathematical or physical Damascus? Jacques Hadamard thought so. Is intuition common sense derived from training, experience, analogy, practice? Is it a hunch, an informed guess, an insight? How does intuition operate? Is it the same as insight or inspiration? Is it the abstracted essence of creativity? Who has mathematical intuition; who does not? We speak constantly of intuition, but is it really there? Or does it have the same ontological (i.e., existential) status as the soul?

There are many mathematicians whose intuition and insight go far beyond common sense and verge on the transcendental or the mystical. In a discussion of the remarkable physicist Richard Feynman, Marc Kac wrote:

> There are two kinds of geniuses, "ordinary" and "magicians." An ordinary genius is a fellow that you or I would be just as good as, if we were only many times better. There is no mystery as to how his mind works. Once we understand what they have done, we feel certain that we, too, could have done it. It is different with the magicians. They are, to use mathematical jargon, in the orthogonal complement of where we are, and the working of their minds is for all intents and purposes incomprehensible. Even after we understand what they have done, the process by which they have done it is completely dark. They seldom, if ever, have students, because they cannot be emulated, and it must be terribly frustrating for a brilliant young mind to cope with the mysterious ways in which the magician's mind works.

Kac probably had the Indian mathematician Srinivasa Ramanujan in mind. Ramanujan (1887–1920), an obscure Brahman clerk, in 1913 submitted to G. H. Hardy, a leading British mathematician, a list of formulas that left Hardy in awe. Some of them, Hardy said later,

> ... fairly blew me away. I had never seen the like! Only a mathematician of the highest class could have written them. They had to be true, for if they were not, no one would have the imagination to invent them.

Ramanujan is considered by many to be a mathematical magician, and, incidentally, has also been proposed as a member of the Asperger Society.

Whatever the terms *mathematical imagination, mathematical intuition, mathematical talent*, and *mathematical genius* mean, they exist as concepts and, while not identical, are related and often conflated. The average person appears to think of mathematical talent as the ability to perform long and complex operations of arithmetic rapidly and correctly. While this talent exists in some people, it has little to do with the talent for mathematical research. Mathematical imagination understands existing theories. It creates new mathematics and, in so doing, often has the courage to go beyond common sense. It reformulates old theories and places them in new contexts. It connects and applies the existing body of mathematics to the exterior world. When this body of work seems inadequate, mathematical imagination devises new mathematics to fill in the gaps or create whole new fields. Mathematical talent is often measured—in the early years at least—by testing; in later years mathematical genius is measured by prizes. I don't know how one can measure mathematical imagination and mathematical intuition except by inferring it from the historic flow of the subject.

On the other hand, intuition is not proof:

> Because intuition turned out to be deceptive in so many instances, and because propositions that had been accounted true by intuition were repeatedly proved false by logic, mathematicians became more and more skeptical of intuition.... Thus, a demand arose for the expulsion of intuitive reasoning and for the complete formalization of mathematics.
>
> —Hans Hahn (1879–1934, mathematician and member of the famed Vienna Circle of Philosophy, the Wienerkreis)

Solomon Feferman, who quotes the above, is much more sympathetic toward mathematical intuition:

> [We should] recognize the ubiquity of intuition in the common experience of teaching and learning mathematics, and the reasons for that: it is essential for motivation for notions and results and to guide ones conceptions via tacit or explicit analogies in the transfer from familiar grounds to unfamiliar terrain. In sum, no less than the absorption of the techniques of systematic, rigorous, logically developed mathematics, intuition is necessary for the understanding of mathematics. Historically, and for the same reasons, it also played an essential role in the development of mathematics. The precise mathematical expression of various parts of our perceptual experience is mediated to begin with by intuitive concepts of point, line, curve, angle, tangent, length, area, volume, etc. These are not uniquely determined in some Platonic heaven. Mathematics models these concepts in more or less rigorous terms (sufficient unto the day) and then interweaves them to form more elaborate models or theories of physical experience as well and purely mathematical theories. The adequacy of explication of the basic concepts can only be tested holistically by the degree to which these theories are successful.
>
> —Mathematical Intuition vs. Mathematical Monsters

A Personal Description of Mathematical Intuition

Unconscious mathematizing is a huge mystery. Mathematical ideas can come into consciousness by surprise, as if by a gift from nowhere evidently from some subconscious process. This is attested by many anecdotes, the most famous involving Poincaré. A mystery calling for explanation.

— Reuben Hersh

If, as Hersh suggests, mathematical intuition remains a mystery (or, as psychiatrist Carl Jung might have said, a manifestation of the "creative unconscious"), I can at least provide a bit of introspection on my thinking process as I worked on two problems. This anecdotal material displays several aspects of intuition.

As I worked along, thinking, remembering, consulting people and texts, learning, writing, computing, manipulating symbols, judging, combining, accepting, rejecting, I was not conscious of something called "my intuition" working on my behalf. I assumed that my experience guided me both consciously and unconsciously and told me that, if I pursued such and such a course, what I was after would probably emerge. When it did I was surprised and satisfied; when it didn't I was disappointed and frustrated. Failures are not reported as often as they should be; they send out warning signals of potholes and roadblocks.

Here are the details (which an uninterested reader may easily skip) of two different personal experiences: (1) my computation of Gaussian integration rules, and (2) my discovery of "quadrature domains."

(1) In the first decade of digital computers, much research was conducted on determining which of the old computational strategies or algorithms were easily coded and were efficient, accurate, and stable. Much research was conducted (and still is conducted) in devising new strategies that satisfy these criteria. One strategy for solving a partial differential equation in compressible fluid theory, and with which I was involved, called for the numerical evaluation of many definite integrals with complex integrands. I recalled that Gauss had devised a method that possessed increased polynomial accuracy over other methods. Its use involved the pre-computation of what are called the Gaussian weight and abscissas. Once these universal numbers are known, they

may be employed for arbitrary integrands. In pre-computer days, these numbers had been computed at the cost of much human labor up to $n = 16$, where n is the number of points used to approximate the value of the integral. I thought it would be a good idea, and useful to the computational community, to extend the determination of the Gaussian abscissas and weight up to at least $n = 48$.

To do so, I had to devise a computational strategy. (I add parenthetically that, while the mathematically unsophisticated may think that when one is confronted with a mathematical problem all one has to do is "give it to the computer," this is far from the case with brand new problems.) The Gaussian abscissas are the roots of the Legendre polynomials. Once these numbers are known, the weights may be computed relatively simply. So my problem was how to compute the roots of the Legendre polynomials of degree up to $n = 48$ accurately and expeditiously. First I had to devise a strategy for computing the polynomials. I knew that the polynomials satisfied a three-term linear recurrence formula with known constant coefficients. What I didn't know was whether the use of this type of recurrence, programmed in the "forward direction," was a numerically stable process, or whether it would blow up. But I persevered.

Now to compute the roots. I had seen graphs of the polynomials of low degree. I noticed that, at their roots, their graphs crossed the axis rather steeply, an indication of high values of the derivatives. This was a good omen for iterative root solvers. But to start an iteration, I needed a fairly close initial approximation to the n roots of the polynomials. Now, I had previously done a fair amount of work that made me quite knowledgeable about the theory of orthogonal polynomials, of which the Legendre polynomials are simply one example. I knew that some deep work of Gabor Szegö in the 1920s and 1930s had provided an asymptotic estimate of the location of the roots of the Legendre polynomials. His book *Orthogonal Polynomials* (1938) had been my bible for a while. I thought that by using Szegö's estimate as a first approximation and then using some form of Newton iteration, perhaps one with cubic convergence to refine the estimate, I would be able to get through.

This, in brief, was my strategy. I did not know whether it would work. I could think of many places and reasons where and why the computation might break down and produce nothing but trash. Nothing daunted, I persisted. I called in the aid of my friend and colleague Philip Rabinowitz, a splendid numerical analyst and a programming whiz. Phil programmed my strategy for the SEAC computer (National Bureau of Standards, Washington, D.C., 1955) in the four-address system that was available on this first-generation machine. As I suggested, Rabinowitz wrote a code for twenty decimal place double-precision calculations. He inserted the code. He pushed the button. SEAC churned away. Eventually, it spat out numbers that independent tests proved to be accurate.

Can the intuitive component of my experience be summed up as memory, knowledge, experience, conjecture, hope, experimentation, fortitude, all followed by a great piece of luck? I never pursued the topic of computing Gaussian weights and abscissas beyond this one experience, but other people followed up on it. The theoretical background guaranteeing the success of the process was filled in. Other, better and quite different, methods and strategies for computing Gaussian integration rules with arbitrary weighting functions were devised and tested.

(2) My second experience is of a somewhat different nature mathematically. In the first experience, I had a definite computational goal. In the second experience, I stumbled across something, quite by accident, quite surprisingly, which I then recognized as an interesting piece of theoretical mathematics. This was mathematical research as exploration. In this type of work one marks out a certain area of interest, brings in information, employs known tools, devises new thrusts, and then sees if anything interesting emerges.

In the early 1970s the following theories rattled around in my mind: orthogonal polynomials; their application to conformal mapping via Gabor Szegö's and Stefan Bergman's kernel functions for two-dimensional domains; Bergman's use of complex conjugate coordinates, $z = x + iy$, $\mathrm{conj}(z) = x - iy$, in the solution of elliptic partial differential equations; and the Hermann Amandus Schwarz reflection principle for analytic arcs. I noticed that given an analytic arc C, the reflection principle could be embodied in the formula $\mathrm{conj}(z) = S(z)$, where the formula serves as the complex conjugate equation of the arc C. The function $S(z)$

is analytic in an open set that contains C. My interest deepened when I considered a form of Green's Theorem written as

$$\iint_B g(z) \, dx \, dy = (1/2i) \int_{\partial B} S(z) \, g(z) \, dz. \qquad (*)$$

Here B is a bounded region of the plane and ∂B is its boundary. For a closed analytic curve, the function $S(z)$ is analytic in a strip that contains ∂B. But $S(z)$ may be extended by analytic continuation, and it may happen that it can be extended (with singularities) into the whole of B.

I was also in possession of a very beautiful formula discovered a few years earlier by Theodore Motzkin and Issai Schoenberg on one of their nocturnal rambles. It expresses the double integral over any triangle T of an analytic function in terms of its values at the three vertices of the triangle. The formula is as follows:

$$\iint_T f''(z) \, dx \, dy = A \, f(z_1) + B \, f(z_2) + C \, f(z_3),$$

where f is analytic over T, z_1, z_2, z_3, are the vertices of T, and A, B and C are three constants that depend only on the z_i explicitly.

I therefore deemed it important to study the formal algebra of $S(z)$, its connection to conformal mapping functions, and its singularity structure. Explicit formulas for $S(z)$ for various closed curves are rare, but a few do exist. I dubbed the function $S(z)$ the *Schwarz Function* for ∂B. This name has stuck, although it conflicts with a totally different object that also goes by Schwarz's name.

If $S(z)$ is meromorphic in B, then $\iint f(z) \, dx \, dy$ can be expressed as a differential operator on f:

$$\iint_B f(z) \, dx \, dy = \Sigma\Sigma \, a_{nk} f^k(z_n).$$

I knew from the work of Szegö and Bergman that, taken over ellipses, the double integral was expressible as a weighted single integral, and as I was working out a variety of algebraic and integral identities, I wondered whether I could find a region B such that

$$\iint_B f(z) \, dx \, dy = (1/2i) \int_I f(x) \, dx, \qquad (**)$$

where the interval $I = [-1, 1]$ was contained in B. In the course of working with Schwarz functions of various singularity structures, I considered the mapping function

$$m(z) = \frac{1}{\pi} \log \left(\frac{1 + az}{1 - az} \right), \qquad (***)$$

which led me to the Schwarz function

$$S(z) = \frac{1}{\pi} \log \left(\frac{1 - e^{\pi} e^{\pi z}}{e^{\pi} - e^{\pi z}} \right), \qquad\qquad (****)$$

which has singularities only at ± 1, $\pm 2ki$, for $k = 0, 1, 2, \ldots$.

And then, lo and behold, and with great surprise to myself, I derived (**) where B is the image of the unit circle under the map (***). The region B turned out to be a symmetric oval-like shape with an explicitly calculable boundary. The result seemed to come to me out of the blue.

Can the intuitive component of this experience be summed up as knowledge, experience, conjecture, analogy, exploration, persistence, all followed by a great piece of luck in which I recognized the significance of what lay before me? There is a saying that "luck favors the prepared mind" and I believe this is an instance of it.

This development—or really, the state of my mathematical concerns in the early 1970s—culminated in a book, *The Schwarz Function and Its Applications*. Having written the book, my curiosity was saturated or exhausted, and though I followed subsequent developments for a while, I never went back to the subject.

What happened afterward? Other people took up the ball and ran it for a touchdown, either as a result of my book or quite independently. The circles, ellipses, ovals, and others, considered in relation to formulas of type (**), were dubbed "quadrature domains," and the theory was pursued by Frank Stenger in Salt Lake City, Harold Shapiro in Stockholm, Makoto Sakai in Tokyo. My results were generalized by these mathematicians, who had much more knowledge and technique at their command than I. Here is what Shapiro wrote recently about the current status of the theory:

Universality [of knowledge] signifies the tendency of mathematical notions to turn up in unexpected places.... Let me cite the (to me very exciting and unexpected) circumstance that quadrature domains, originally discovered in connection with an extremal problem in conformal mapping, turned out to relate to fluid mechanics and inverse problems of gravitation. Moreover, investigations in operator theory involving aspects of subnormal and hyponormal operators have also led, surprisingly, to quadrature domains.

Further Reading

P. Davis and P. Rabinowitz. "Abscissas and Weights for Gaussian Quadratures of High Order." *J Research, National Bureau of Standards*, Vol. 56, 1956.

For Gauss quadrature after the work just cited, see, e.g., the next two references.

G. H. Golub and J. H. Welsh. "Calculation of Gauss Quadrature Rules." *Math. Comput.*, Vol. 23, 1969.

Walter Gautschi. "A Survey of Gauss–Christoffel Quadratura Formulae." In the Proceedings of the Christoffel Symposium, Aachen, 1979.

Reuben Hersh. "Does 'Existence' Matter? Mathematical Practice as a Scientific Problem." In *Current Issues in the Philosophy of Mathematics from the Viewpoint of Mathematicians and Teachers of Mathematics,* Bonnie Gold and Roger Simons, eds. Mathematical Association of America, 2006.

P. N. Swarztrauber. "Computing the Points and Weights for Gauss-Legendre Quadrature." *SIAM Journal on Scientific Computing,* Vol. 24, 2002.

Philip J. Davis. *The Schwarz Function and Its Applications*. Carus Monographs, Vol. 17, Mathematical Association of America, 1974.

Philip J. Davis. *The Education of a Mathematician*. A K Peters, 2000.

Philip J. Davis and Reuben Hersh. *The Mathematical Experience*. Birkhäuser Boston, 1980.

Solomon Feferman. "Mathematical Intuition vs. Mathematical Monsters." *Synthèse*, Vol. 125, 2000.

Jacques Hadamard. "An Essay on the Psychology of Invention in the Mathematical Field." Princeton University Press, 1945.

G. H. Hardy. *Ramanujan: 12 Lectures on Subjects Suggested by His Life and Work.* Cambridge University Press, 1940.

Roger Penrose. *The Emperor's New Mind*. Oxford University Press, 1989.

Makoto Sakai. *Quadrature Domains*. Lecture Notes in Mathematics, Vol. 934, Springer-Verlag, 1982.

Harold S. Shapiro. *The Schwarz Function and Its Generalization to Higher Dimensions*. Wiley, 1992.

6

Are People Hard Wired to Do Mathematics?

> I was not driven to become a scientist by any craving to understand the mysteries of the universe. I never sat and thought deep thoughts. I never had any ambition to discover new elements or cure diseases. My strong suit was always mathematics; I just enjoyed calculating, and I fell in love with numbers. Science was exciting because it was full of things I could calculate.
>
> —Freeman J. Dyson

The vague expression "hard wired" has entered into common parlance. But what does it mean? In some discussions, "hard wired" seems to imply a genetic inheritance, nature, as opposed to nurture; in other discussions, the expression relates to brain physiology, implying that the various connections in the brain correspond to a wiring diagram.

If we humans are hard wired, not all of us appear to be hard wired to the same degree or in the same way. Are people hard wired to do mathematics? In a trivial sense, this is certainly the case. When people do mathematics, they are using their brains and not their elbows, and since people *are* capable of doing mathematics, the wiring of their brains (whatever that means) is sufficient to support mathematical thinking. Closer to brain physiology, there is evidence arriving, principally from new brain scan techniques, that this is indeed the case. Further questions: are some hard wired *against* doing complicated mathematics? Is there a gender difference in the "mathematical wiring"?

At the far end of creativity, consider Asperger's syndrome (named for Hans Asperger, an Austrian pediatrician, 1906–1980), about which there has been much speculation. This condition is characterized by poor social interactions, repetitive behavior patterns, and numerous eccentricities, together with specific talents, often mathematical. The mathematical talent displayed by people with this syndrome can be considerable. But when we attach a clinical name to a certain type of personality, we should think of it as part of a continuum of personalities that, along many dimensions, correspond to many different mental modes of operation.

The physicist David Park, who has had close experience with autism, has written to me:

> Normality, Asperger's, and autism are all regions in an n-dimensional space. Persistence may be a relevant dimension, energy may be another, imagination another, and surely there are "smarts." But pure, unadulterated smarts can't create the Newtons of the world. And how large is the dimension n? I don't know, but it is large enough to accommodate these four (and doubtless other named conditions) in the same space. Does the space have a metric? I don't know, but there may be some sense in which some Asperger people are closer to normality and others closer to autism. This statement makes sense to me but with so many variables, each, perhaps, defined in behavioral terms, I don't really know how to formulate it.

I myself have witnessed all these dimensions in the creative mathematicians I have known, but I wonder just what pure unadulterated "smarts" would consist of.

Is a touch of Asperger's necessary for achievement in mathematics? Asperger himself believed so, and so do numerous professional mathematicians. If one looks for whatever makes an artist or a mathematician or a whatever, it may turn out to be as much a personality type as a specific hard-wired aptitude.

Sir Isaac Newton was asked how he had solved so many problems, and he answered, "By always thinking upon them." Quite apart from those who are diagnosed with Asperger's syndrome, many people become engrossed by one question or problem and, over long periods of time, think about it obsessively.

Neuropsychologist Stanislas Dehaene points out that brain studies have located certain elementary mathematical operations in specific areas of the human brain. On the other hand, he cautions,

> In the final analysis, where does mathematical talent come from? ... Genes probably play a part. But by themselves they cannot supply the blueprint

of a phrenological bump for mathematics. At best, together with several other biological factors, perhaps including precocious exposure to sex hormones, genes may minimally bias cerebral organization to aid the acquisition of numerical and spatial representations.

Consider how spatial abilities in infants increase rapidly as they grow older, and then think of how many lines of mathematical code it would take to have a robot simulate such abilities. Probably hundreds of thousands of lines, and, truth to tell, no one has any idea at all how to do this. So in this sense, one might say that an infant is solving difficult mathematical problems unconsciously by employing its wired-in capabilities. Even animals appear to be hard wired for some very elementary mathematical abilities.

While the performance of arithmetic operations (as well as most categories of cognitive activity) appears to be localized in the brain, the performance calls into play vast neural capacities. Is, therefore, the human brain a digital computer? The evidence is that it possesses some features of an analog-digital computer, but this is hardly the whole story of what the brain "is."

What, then, makes a professional mathematician? Facility with numbers is not sufficient. But what else is needed? Isn't curiosity at the heart of most scientific effort? Isn't there the enjoyment of extreme mental effort even when, as usual, it does not lay golden eggs? Because, even then, one gets a kick out of it, as Freeman Dyson implies.

Further Reading

Stanislas Dehaene. *The Number Sense: How the Mind Creates Mathematics.* Oxford University Press, 1997.

Keith Devlin. *The Math Instinct: Why You're a Mathematical Genius (Along with Lobsters, Birds, Cats and Dogs).* Thunder's Mouth Press, 2005.

Apostolos Doxiadis. *Uncle Petros and Goldbach's Conjecture.* Bloomsbury, 2000.

Freeman J. Dyson. "Member of the Club." In *Curious Minds: How a Child Becomes a Scientist,* John Brockman, ed. Pantheon, 2004.

Ioan James. "Autism in Mathematics." *Mathematical Intelligencer,* Vol. 25, No. 4, Fall 2003.

7 Why Counting Is Impossible

A mathematician once declared: There are three types of
mathematicians, those who can count and those who can't.
—Old Joke

What does it mean "to count"? How does one understand what exactly
it means to be counted? How is counting to be done? What potholes
or roadblocks are encountered on the way? Of what use is counting?
Webster says that to count is "to indicate or name by units or groups so
as to find the total number of units involved," but, as is frequently the
case, this dictionary definition hardly clarifies the situation.

At the dawn of mathematics, counting was a physical act done by
humans. Notches were cut in a stick; tallies were drawn in the sand,
groups of four crossed to make five. This led to numeric symbols, and
these, in turn, led to arithmetic rules that might be termed the grammar
of arithmetic. Probably the earliest records of formalized mathematics
begin with counting and numeration. Sumerian tablets (c. 1500 BC) give
multiplication tables and reciprocals in base 60. These tables already
go far beyond mere counting. According to cuneiformist Alice Slotsky,
there was already a tendency to "pad" the numbers so that the kings
would come off looking great and strong. It seems that hanky-panky
with regard to counting has been around from the very beginning!

More generally, Helen de Cruz has written:

Cross-culturally, humans count collections of objects larger than three
or four items by putting them into a one-to-one correspondence with a
symbolic unit (e.g., words for numbers, fingers or other parts of the body,
notches on a tally). This universal cultural phenomenon sets *Homo sapi-
ens* apart as the only animal that can count large collections of objects

accurately. I will argue that both the ability to estimate and compare numerosities, and to accurately count collections of four or less objects (*subitization*) is hard-wired in the human brain, and that we share this ability with many other animal species. However, humans use tangible and intangible symbols (so-called *artificial memory systems*) to count larger collections of objects. I will show that this universal (cultural) human feature is part of a key adaptation in human evolution, which arose relatively late, about 50,000–40,000 BC in Africa during the Middle Stone Age [to] Late Stone Age transition.

It became clear early on that when the objects to be counted were too numerous—say, two or three thousand jelly beans in a jar—the physical process cannot be carried out with anything like the absolute fidelity required by a Platonic vision of mathematics. This vision asserts that there is one answer and only one answer to how many jelly beans there are in the jar. A further vision in the mathematical trade says that we can count 1, 2, 3, ... going on "forever" and do it with perfect fidelity. Common sense says this is nonsense, but pure mathematics puts us in an idealistic frame of mind. Cosmologists may talk about "the big crunch," the reverse of "the big bang," in which the cosmos and all therein disappears completely, but the world of mathematical ideas abides in a platonic heaven counting away: 101,000; 101,000 + 1; ... (to paraphrase Goethe, *Das ewig Ideal zieht uns hinan*).

Cognitive psychologists tell us that the eye–brain connection cannot immediately grasp multiplicities beyond seven or eight displayed objects.[1] Experience tells us that successive counts of large numbers (even when made by different people) may yield different results. This is called fuzzy math. The term "fuzzy" has a technical sense, but I'm using it to mean processes involving mathematics that yield several different answers where a single answer is required. There is a common rule of thumb to defuzz: Do it over and over again until two successive counts yield identical results. Then stop, and adopt the common value.

---

1. Recollection of long strings of numbers is another issue. It has been asserted that more than ten digits (e.g., area code plus phone number) exceeds most people's comfortable memorization ability. In payment of a periodically sent bill, customers of a gas company are asked to write their account number on their checks for "security reasons." I wonder how often customers make an error in simply copying the number.

This is the rule used by the old-time grocer who had a pencil behind his ear and moistened it on the tip of his tongue. He first added the column of prices going down, then he added the column going up. This is not unlike the exit strategy for computations in numerical analysis: Do it once, do it twice, or do it over and over until the difference between successive computations is, we hope, less than an allowed margin of error, often designated by epsilon. Numerical analysts then theorize about the size of epsilon.

Here is another example of the difficulty of counting—entirely within the realm of computer technology. MATLAB™ is a software package used for scientific computation. In an earlier version of the software, it was possible after an extensive computation was done to ask for the number of flops that the computation required (flops = the number of floating point operations). This number was regarded by some analysts as providing a measure of the complexity or the efficiency of the program performing the computation.

Later versions of MATLAB omitted this feature. I asked Cleve Moler, co-founder of the company that created MATLAB, why this was the case. His answer, briefly, was

> This popular MATLAB feature was a casualty of the introduction of LAPACK. The trouble is that LAPACK and, especially, optimized BLAS [where most of the floating point operations are now done] have no provision for keeping track of counts of floating point operations. Unfortunately, it would require more extensive software modifications than we are able to make. There might also be some degradation of performance.

Moler went on to say that there may now be little reason to count flops: "With modern computer architectures, floating-point operations are no longer the dominant factor in execution speed. Memory references and cache usage are most important."

Consider the mathematical/political/legal embroilment known as the 2000 Presidential Election. In the case of that election, we were asked to count to one hundred million (100,000,000) or beyond. From the start to the finish, thousands and thousands of people and machines were involved in the process. Should the results have been checked? Certainly, but how? What is the probability that simple counting errors will cancel out? What is the probability that gross fraud will not? Who will check the checkers? And who will check the check-checkers?

The controversy was over simple counting, an instance of applied mathematics. The turmoil surrounding the election involved counting, recounting, butterfly ballots, the definition of which ballots were to be counted, legal challenges, court decisions, and further appeals. The nation desperately looked for an "exit strategy," and the exit strategies proposed came under litigation themselves. The tightness of the race reminded me of the unbelievable, seemingly nonsensical, "butterfly phenomenon" of chaos theory. But we have now experienced this phenomenon: a butterfly flapped its wings in Florida, and the whole country was affected.

The controversy concerning the manner in which votes are counted has deepened beyond paper ballots. On-line voting has attractive features both to the public and to the companies that have devised the soft- and hardware (and have made much money thereby).

The opponents of on-line voting have come up with many reasons for not installing such systems: terminal usage may not be available to the poor or uneducated; the secrecy of the ballot might be compromised; multiple voting by an individual and vote theft on a large scale could easily be arranged by hackers; viruses that change votes might be installed in computer terminals. Even worse, the companies or governments that control network access might tamper with votes in an essentially undetectable way. It's possible that an open device—that is, one where all the layers of hardware and software are open for inspection—would suffice, but a paper trail, which many people demand, is much easier. In any case, the people who build voting machines seem to have no interest in building an open machine, and the side-by-side existence of such a machine and a paper trail would pose its own serious problems.

The convenience of on-line voting cannot be denied. It is used for opinion polls, proxy voting by company shareholders, etc. Since there is no flawless system of voting, my prediction is that on-line political voting will gradually become standard procedure. And there is no system of declaring a winner—including systems such as proportional representation, the electoral system, even majority rule[2]—that doesn't have substantial flaws.

Where we once used tally sticks, we now use computing machinery to tally and reckon. Zillions of such operations are carried out and accepted without question every day; but with respect to some critical economic, social, or political issues, such as elections, some people, dreaming of a simpler age, may cry out: "Do it by hand. Hand tallies are more correct." To them, I say, no way, José: The computer is here to stay.

In certain instances, the computer may err. The computer may have been programmed erroneously, whether deliberately or inadvertently, or it may have malfunctioned. How does one check the programming or physical operation of the computer? Do the tally over again, preferably independently and on a different computer. For increased fidelity in a complex mathematical problem, it may be possible to check the output in a special case where one happens to know a "closed form answer" in terms of standard, well-known functions.

Consider the problem of counting the population of the United States, recently the subject of litigation. The basic question of who should be counted is moot. For example, should a prisoner be counted in the prison district or where his or her home is? It could make a difference both politically and financially. Mathematics proposes a number of counting schemes, each method of which has plusses and minuses. Conflicting evidence was presented to the courts by qualified statisticians in suits relating to how U.S. Census 2000 counts were to be made. Should the counts be actual "nose" counts or by statistical sampling? The ultimate choice is not mathematical; it is political and legal. Everyone agrees that simple nose-counting results in undercounts of certain groups. On December 5, 2002, the Census Bureau, under court order, supplied two sets of figures (fuzzy math!): simple counting and counting augmented by statistical sampling. By a unanimous decision of the U.S. Supreme

---

2. In the 2004 gubernatorial race in the State of Washington, after counts, challenges and recounts, Democrat Christine Gregoire was declared the winner over Republican Dino Rossi on the basis of a plurality of 129 votes out of more than two and a half million. Would a plurality of one vote have sufficed? The statisticians are tearing their hair out. So what is the definition of a majority?

Court, only the simple count will be allowed to determine congressional reapportionments. And speaking of reapportionments, the gerrymanders that go on are now creating a new branch of applied mathematics that might be called "political geometry" and have also been subject to Supreme Court review.

The pursuit of mathematics depends on a high level of often-unacknowledged trust and faith. If this is true for mathematics, it is most certainly true for human dealings. Trust has now been corroded on both sides of the political aisle. The Law was not able to anticipate or deal with all the consequences of an unavoidable amount of fuzzy math. Hence the vibratory back-and-forths of legal rulings that might be called political Parkinsonianism. Shortly after the Election 2000 crisis deepened, former President Bill Clinton quipped, "The public has spoken. But what has it actually said?" The mathematicians and their machines have spoken, but what have they said?

Counting relates to the intuitive mathematics of the real world in an exceedingly complex way. It is the simplest and the most fundamental operation in mathematics, but it is a human impossibility when the numbers are sufficiently large and when absolute and irrefutable fidelity is expected. The value of a large corporation is not determined by counting, among other things, the number of light bulbs in its factories and offices. There are standardized accounting procedures set up by the profession of accounting for arriving at a value. Different procedures may yield different values, and the value accepted in the commercial world may depend on a variety of considerations. For instance, should the company's pension fund be listed as an asset? Accountants disagree on this point. Released numbers are often aimed at certain audiences and are "massaged" to that end. Corporations sometimes release several balance sheets. Thus, for economic and social questions, the accepted "count" may depend ultimately on litigation and judicial decisions, and these issues make up the literature of a relatively new branch of applied mathematics known as jurimetrics.

Thus, applied mathematics abounds with fuzzy math, and we have learned to deal with it. But what about pure mathematics? Surely there is no fuzz there, common sense assures you. But let me speak of a recent personal experience.

There is an incorrect mathematical statement in *The Mathematical Experience* that has taken two decades to surface. My co-author Reuben Hersh found it recently on a website titled Bertleson's Number. The function $\pi(x)$ denotes (or counts) the number of primes less than x. On page 175 of our book, we listed $\pi(10^9)$ (Bertleson's Number) as 50,847,478, while on page 213 we listed it as 50,847,534. Both numbers may be wrong, but both cannot be right. We did not compute the number ourselves: In the process of writing, we copied it from two "reliable" sources.

The first value (50,847,478) came from G. H. Hardy and E. M. Wright's famous *An Introduction to the Theory of Numbers*. The second value (50,847,534) is cited by the website author as the correct value. According to my philosophy, however, there is no way of determining with absolute fidelity the value of $\pi(10^9)$. Luckily, we have many methods for increasing the probability that a given answer is correct. As mentioned, several independent people, preferably using different methods and different computer software, may redetermine the value. Errors can be corrected, but how many independent redeterminations are sufficient before one goes into print? We didn't realize that we had built in a contradiction. And, actually, the exact eight-figure value of $\pi(10^9)$ was of little importance to us. We had other rhetorical and philosophical fish to fry.

The literature of pure mathematics is so vast that there are many errors in it, and not just counting errors. Most errors lie undiscovered until something important rests on one of them and someone investigates. People have compiled lists of errors, the lists themselves containing errors. Imre Lakatos' classic book *Proofs and Refutations* presents the history of the Euler–Poincaré theorem as a comedy of errors. In my article *Fidelity in Mathematical Discourse* I gave a taxonomy of types of errors that can occur.

One characteristic of applied mathematics as it pursues utility and calls action into play is that it must have an exit strategy that operates within reasonable human time. Pure mathematics cannot have an exit strategy. Its statements are forever open for correction, for improvement, for reinterpretation and development.

While we may not be able to count with an idealized absolute fidelity, we can make useful estimates, and we are forever making such estimates. If, for some reason, we want to know the gross national product of the United States, the population of the world, the number of reindeer in Lapland, the number of galaxies in the cosmos, the number of people who rate a certain movie star highly, or the number of blood cells in the body, we cannot find the number by direct counting. But there are many ways of arriving at decent, good-enough, or ball-park estimates. The scientific and technological value and even the legal status of such estimates have to be determined on a case-by-case basis. Here are a few of the ways to get these good-enough estimates.

1. Mechanical counters, as in turnstiles or the traffic counters placed on highways. Do these give exact numbers? In principle, yes. In practice? People have been known to jump over turnstiles.

2. How many people thronged together in Times Square on New Year's Eve? The NYPD will give you an estimate. I wonder how they arrived at the number: by counting heads in photos? Regarding political rallies, in general the police give a smaller number than the organizers do, and these estimates often differ by a factor of 5 or more.

3. Arithmetical estimating by "eyeballing" or rounding. For example, Estimate $2,345,678 \times 6715$. Rounding the first number down and the second number up changes the problem: Find $2,000,000 \times 7000$. A thousand million is a billion, and $2 \times 7 = 14$, thus the product is about 14 billion, or $14,000,000,000$. Exact answer from my computer: $15,751,227,770$, so the rounding estimate was not far off.

    Some arithmetic may be difficult to eyeball, however. Consider the following problem:

    Estimate $(6.78 \times 1459) - (7.68 \times 1280)$.

    It's possible that a "lightning calculator," i.e., a person who can do large computations mentally, may be able to give you an answer after eyeballing this problem. But I'd go for pencil and paper or, better yet, I'd run to my computer.

4. "Counting without counting"; this is what the branch of mathematics known as combinatorics is all about. Examples:

    How many strings of length n, consisting either of 0 or 1, are there? Answer: 2^n.

    In how many different ways can we select 2 items from a pile of n items? Answer: $n(n-1)/2$ ways.

In how many different ways can one put $(a + b)$ objects into two boxes so that the first box holds a objects and the second box holds b objects?

Answer: $\frac{(a+b)!}{a!b!}$ ways.

In how many different ways can you represent the positive integer n by means of sums of positive integers without regard to order? This number is designated as $p(n)$, the partition function. Example: $5 = 1 + 4 = 2 + 3 = 1 + 1 + 3 = 1 + 2 + 2 = 1 + 1 + 1 + 2 = 1 + 1 + 1 + 1 + 1$. Thus $p(5) = 7$. The formulas for and the theory of the partition function are by no means simple.

5. Derivations using common observations, conjectures, rules of thumb, and mathematical or physical formulas. An ancient and oft-cited example of this can be found in Archimedes' book *The Sand Reckoner* (c. 225 BC) Addressing King Gelon, Archimedes states that the number of grains of sand to fill the universe is not infinite and that he can give a number for it. Estimating the size of a grain of sand and using the estimate given by the astronomer Aristarchus of Samos for the size of the sphere of the fixed stars, he arrives (in our notation) at 8×10^{63} grains. (Archimedes certainly knew the formula for the volume of a sphere.)

6. Statistical sampling followed by a statistical analysis. Sampling is given frequent public notice in social and political polling. But what meaning or credence, personal or otherwise, can we really give to statements such as "Candidate X comes in at 38% with an error margin of ±5%."

Despite all these difficulties and uncertainties we are constantly counting. And then we add, subtract, and mathematically manipulate our counts. I shall elaborate further on this in later essays.

Further Reading

Kwesi Addae and Jason Mark discuss their involvement in the Fair Election program, which has run election monitoring in ten countries, on their website (http://fairelection.us/index.htm). With trust in the electoral process diminished in the United States since the 2000 election, and with new concerns about electronic voting, Fair Election has brought a multinational, independent, nonpartisan, and nongovernmental team of skilled election monitors to examine and report on the U.S. electoral process within the framework of international election standards. Fair Election is a project of Global Exchange (http://www.globalexchange.org/).

Mark Burgin. *Diophantine and Non-Diophantine Arithmetics* (http://arxiv.org/ftp/math/papers/0108/0108149.pdf).

I. Bernard Cohen. *The Triumph of Numbers: How Counting Shaped Modern Life.* W. W. Norton, 2005.

Philip J. Davis. "Fidelity in Mathematical Discourse." *Amer. Math. Monthly,* March, 1972.

Philip J. Davis. "Count, Recount, and Fuzzy Math." *SIAM News,* Vol. 34, No. 1, 2001.

Philip J. Davis and Reuben Hersh. *The Mathematical Experience.* Birkhäuser Boston, 1980. Numerous foreign language translations available.

Helen de Cruz. "Why Humans Can Count Large Collections of Objects Accurately." In Workshop on the Social Dimensions of Mathematics Brussels University (VUB), Belgium, December, 2003.

Imre Lakatos. *Proofs and Refutations.* Cambridge University Press, 1976.

Cleve Moler, e-correspondence.

Lynn Arthur Steen. *Achieving Quantitative Literacy: An Urgent Challenge for Higher Education.* Mathematical Association of America, 2004.

8
Quantification in Today's World

Our age is awash in rank-orderings and quantifications of all sorts. In recent years, I have seen quantified the quality of a restaurant, the aesthetic content of a work of art, the consonances and dissonances in a musical composition, the degree of compatibility of two young lovers, the prestige value of checking into the Plaza Hotel, the degree of belief in a statement, the degree of risk in an enterprise, and the competitive utility of a product. I have seen quantified the performance of an Olympic skater, the productivity of a nation, the appropriateness of job J for individual I, the caliber of a university department of mathematics, the tendency of Nation A to go to war with Nation B, the degree of acceptability of a given prose style, the condition of the world's poor, and the excellence of a movie. I have even heard of the mathematization of beauty as a computer aid to cosmetic surgery (see Steven G. Krantz, in the Further Reading section). Readers can easily extend these lists. While such quantifications often embody a consensus, reached by averaging the results of a salad bar of responses or the opinions of a panel of experts, they do not as yet command firm respect or agreement as to their appropriateness or utility.

The questions of what can be quantified usefully, what can be rank-ordered, and how one goes about doing it, can be answered only historically. The equations of physics are phrased in terms of quantification. The relation between physics and the quantifiable is an intimate one, and its necessity is now considered absolute:

The demand for quantitativeness in physics seems to mean that every specific distinction, characterization, or determination of a state of a physical object and the transmission of specific knowledge, must ultimately be

expressible in terms of real numbers, either single numbers or groupings
of numbers, whether such numbers be given "intensively" through the
medium of formulae or "extensively" through the medium of tabulation,
graphs or charts.

—Salomon Bochner

As the great physicist James Clerk Maxwell said, if something is
expressible in numbers, then it is understandable and potentially scientific. The psychologist E. L. Thorndike made a more sweeping assertion: "What exists, exists in some amount." Hence, all becomes grist for
the mathematical mills, which can grind quantities with ease and then
have them baked into "scientific" bread. We get used to this. But all is
not that easy. As the mathematician André Weil remarked, mathematics
provides a palette of objects and operations from which we must pick
and choose with care.

How can musical consonances and dissonances be described quanti-
tatively? A reader might believe that since knowledge of a simple relation between the length of a vibrating string and the harmonic sound it
produces dates back to the ancients, this problem was solved long ago.
Not so. Would you believe that the famous astronomer Johannes Kepler
(1571–1630) had a theory of musical consonances involving the particular polygons that can be constructed by rule and compass means? H. F.
Cohen says that the consonance problem is not yet solved; theories are
still being advanced.

Move away from physics, and consider concepts like happiness and
pleasure. Statistician F. Y. Edgeworth (1845–1926) believed that pleasure could be quantified, and that, for example, my pleasure could be
added to your pleasure to arrive at something significant. In this way,
a "hedonic calculus" could be set up. He asserted later that utility and
degrees of belief could be measured, and in this way came to econometrics and rediscovered one of the standard interpretations of probability.

Consider the question of the IQ (intelligence quotient). For many
years, intelligence has been measured or described by a single number.
Psychologists now tell us that an individual may possess many "intel-
ligences," perhaps as many as eight, so that quantification of intelligence, if possible at all, must be done vectorially. Let's assume that
this has been done. Since an eight-dimensional number loses the rela-

tionships of greater than and less than, and since the space of eight dimensions is far richer than the one-dimensional space of traditional IQ, we can rightfully ask what conceivable use we could make of eight-dimensional intelligence quotients, and how we could go about validating such usages. Yet, there is a thought that in selecting people for a given purpose, we might be able to use only those dimensions that are relevant to that task.

Another example, not necessarily more dubious than IQ, is the notion of freedom. In 1990–1991, the United Nations promulgated a mathematized notion of freedom and, on this basis, ranked countries as to their levels of freedom. Entering into the "freedom quotient" were such considerations as freedom of women to have abortions and freedom to engage in homosexual practices. Later, due to adverse criticism, they discontinued this index.

We could easily compile, I feel sure, an Index of Prohibited Indices.

Measuring the Unmeasurable: The Subjective Becomes Quantifiable

Whatever is quantitative carries with it the cachet of the scientific, the objective. What is subjective is merely your experience, your perception, your opinion, and is not necessarily mine and may not be expressive of the "true" situation.

It is the easiest thing in the world to pass, or to claim to have passed, from the subjective to the quantitative. How? One of the ways, surely the crudest, but one that is extremely popular, is merely to ask people to rate their perceptions on a scale:

- On a scale from 0 to 10, how much do you like broccoli?
- On a scale from 0 to 10, how well do you think the President is doing his job?
- On a scale from 0 to 10, indicate the degree of social alienation caused by the suburbanization of Crescendo, Pennsylvania.

There are no difficulties in eliciting answers to such questions: The public is delighted to answer. Once we have numbers, we can operate on the numbers mathematically and automatically and derive consequences or cause certain actions to be initiated.

Much thought has gone into constructing scales and into considering the invariants deemed necessary. There is a branch of mathematics known as abstract measurement theory, or metrology, begun in 1887 with Hermann v. Helmholtz. It is pursued largely by mathematicians and psychologists. The theory is both highly abstract and philosophically moot.

If I come down heavily against the transition from the subjective to the quantifiable and thence to the objective, let me point out that in some "hard" sciences, the preference for the objective over the subjective has had its ups and downs. What, for example, could be considered more objective than a photograph? Yet in astronomy and in anatomy, reliance on the "evidence" of drawings, often done by skilled artists, has given way to reliance on photographs, and then, after some years back to drawings.

Quantity versus Quality

According to some philosophers, quality and quantity are independent attributes. Once perceived, moreover, a quality cannot be reduced to more primitive concepts. While dialecticians observe the change in quality wrought by quantity, scientific and mathematical minds tend to see the matter more intensely. Consider, for example, how the quality of driving along a freeway varies with the density of the traffic. Quality is seen as depending on and subordinate to quantity. As Lord Rutherford, Nobelist in physics, once put it: "The qualitative is nothing but poor quantitative."

What do we mean when we speak of the quality of a piece of cloth? We could be referring to its thickness, tightness of weave, stretchability, durability, or shrinkage after washing; all of these characteristics are measurable. Or we could be talking about its "hand," how it feels when you touch it; this characteristic not measurable. We use mathematics to convert some questions of quality to measurable and easily processed quantities.

In some mathematical situations, particularly in the pre-electronic computer days, qualitative statements were more easily arrived at than quantitative statements. In the last century, there has been an emphasis on qualitative theories and a denigration of the quantitative. (I'm thinking here of qualitative differential equations.) More generally, some believe that the Bourbaki treatment of mathematics (c. 1940) was based on structures and on qualities rather than on quantities. The digital computer has turned this around, and the newly developed and extensive theories and practices of pattern recognition have merged quality and quantity in a way that would please Lord Rutherford.

Further Reading

Salomon Bochner. "The Role of Mathematics in the Rise of Science." In *The Dictionary of the History of Ideas,* Scribners, 1974.

H. F. Cohen. *The Quantization of Music.* Kluwer Academic Publishers, 1984.

Alfred W. Crosby. *The Measure of Reality: Quantification in Western Society 1250–1600.* Cambridge University Press, 1997.

Philip J. Davis. "The Unbearable Objectivity of the Positive Integers." *SIAM News,* Vol. 32, June 1999.

Francis Y. Edgeworth. *Mathematical Psychics: An Essay on the Application of Mathematics to the Moral Sciences.* C. K. Paul, 1881.

Peter Galison. "The Birth and Death of Mechanical Objectivity." In *Picturing Science, Producing Art,* Caroline A. Jones, Peter L. Galison, Amy E. Slaton, eds. Routledge, 1997.

Hermann von Helmholtz. *Zählen und Messen erkenntnistheoretisch betrachtet* (*Counting and measuring seen from the point of view of epistemology*), 1887. Reproduced in *Hermann v. Helmholz—philosophisch Aufsätze und Vorträge,* H. Wollgast Hörz, ed., Akademie Verlag, Berlin, 1971.

Steven G. Krantz. "Conformal Mappings." *American Scientist,* September, 1999.

Theodore M. Porter. *Trust in Numbers: The Pursuit of Objectivity in Science and Public Life.* Princeton University Press, 1995.

When Should One Add Two Numbers?

Recently a book came across my desk for review that contained about seventy separate articles written by a variety of authors. It occurred to me to ask, "How on earth could I, or anyone, review such a thing?" I finally thought of a way. Using a random number generator, I would select one or two sentences at random from each article and paste them together. Figuring about 15 words from each of the selected sentences, and adding the lengths, I would produce an article of about 1500 words, which is about the average length of the reviews I write, and turn the result in to my editor. Although sampling is a well-studied and frequently employed device, although randomness has not infrequently been used in modern art, I think the reader will agree that the method I've just proposed, which emphasizes procedure at the expense of meaning, is totally mindless.

Now consider a simpler process: When is it sensible, in the physical world, to add, to concatenate, to paste together, to include? Can one add apples and oranges? Sugar and flour? In a recipe, one surely can. Can one add a list of the dogs in a vet's kennel to a list of Shakespeare's plays? Would it make sense? Why not?

The abstract properties of simple addition are well known and have been studied at length: $a + b = b + a$; $a + (b + c) = (a + b) + c$, etc., and these have been interpreted in cognitive terms. There is, however, no simple all-inclusive meta-rule as to the real-world situations in which these properties hold. Hence there can be, in advance, no general description of the situations in which mathematical addition is called for and is appropriate.

The word "add" is used in diverse scenarios. From the viewpoint of cooking procedures, a cup of milk added to a tablespoon of flour is not the same as a tablespoon of flour added to a cup of milk. In a given situation, how do we know whether mathematics is called for? And if it is called for, under what circumstances is simple addition appropriate?

Two grapefruit together with an additional three grapefruit undoubtedly make five grapefruit in the fruit store. Three electoral votes in Nevada plus 26 in Florida undoubtedly make 29 electoral votes. But the order in which the eastern votes are added and are made known to the public may affect the voters in Hawaii. One cup of dirt added to one cup of water does not make two cups of mud. One million miles an hour added to one million miles an hour do not make—relativistically speaking—two million miles an hour. There used to be a standard word problem in school: "If one pipe can fill a tank in five minutes and another pipe can fill it in ten minutes, will it take fifteen minutes to fill the tank?" Well, the tank will certainly be full after fifteen minutes, but we might have to call in the plumber to wipe up. If my wife and I had had eight children instead of four, then would our pleasure have been doubled? What about our expenses? If one pill is good for a medical condition and two are better, three pills may kill you.

A quantitative situation in which simple addition is appropriate is often called "linear." We are all infected by "linear-itis," the tendency to add things together when in doubt. Adding is easy, and it often works, but most human concerns are non-linear. This is because humanity is encompassed by limits. The Weber-Fechner Principle, which provides an approximate sensory response S for a given intensity of stimulus R, $S = k \log R$, is only one of many instances of the mathematization of human nonlinearities.

The further you go away from zero the more closely you approach a limit for linearity; beyond that limit, non-linearity sets in. There is no generic way of recognizing and hence formalizing what is and what isn't a linear phenomenon. Does this mean that we can never know when to add? No, of course not. Experience guides us, but it can also delude us. Adding, often mindless adding, is ubiquitous. Consider the point systems used in college admissions, applications for mortgages, insurance policies, psychological tests, and much more. If you are over sixty, two points; if you have lived at your current address for more than ten years, one point; if your SAT score is x, then you've earned $x/100$ points; if you have a smoke alarm, one point. Add up all the points. This mathematics makes for easy decisions, and these decisions are then said to be objective. I have seen this system employed by a university selection committee in considering candidates for an open faculty position. So many points for teaching ability, so many points for papers published, so many points for contracts and grants, etc.

Children are taught how to add and subtract, and they develop rules of thumb when confronted with word problems. Stella Baruk reports, in her revealing book *L'Age du Capitaine*, that when a word problem presents several small numbers, children add them. When the word problem presents two numbers, one of which is substantially greater than the other, children subtract them. This is the common sense of taking tests, as confirmed by their experience.

What does this tell us about children? What does it tell us about tests? The children have invoked a context-free situation to solve a specific word problem. Isn't that precisely what pure mathematics does? In truth, we cannot formulate a blanket rule that will inform school children (or ourselves) when it is appropriate to add. And if this is the case for "mere" addition, it is much more the case for the numerous complex numerical operations that mathematics has devised. This is what makes applied mathematics so difficult.

Further Reading

Stella Baruk. *L'Age du Capitaine. De l'erreur en mathematiques*. Editions du Seuil, Paris, 1985.

Mark Burgin. *Diophantine and Non-Diophantine Arithmetics: Operations with Numbers in Science and Everyday Life* (http://arxiv.org/ftp/math/papers/0108/0108149.pdf).

10 Category Dilemmas

When, in real life, we have to fit events or phenomena into sharply defined categories, we must draw the lines between them somewhere. This is frequently done after some deliberation, and the demarcation is often based on mathematical criteria. The lines thus drawn are often in conflict with common sense. And where does one draw the line when there is no clear separation?

An ancient problem associated with categories and quantification is known as the sorites paradox. The word "sorites" comes from the Greek word σορoσ which means "heap," and the paradox goes back as far as Eubulides of Miletus (4th century BC), the man who also formulated the liar paradox, often stated as "This sentence is false."

The sorites paradox is often given in the following form: One grain of wheat does not constitute a heap. Adding one grain of wheat to what is not a heap won't turn it into a heap. By induction, 100,000 grains of wheat would not make a heap. Yet, we know that there are heaps of grain. Where do you draw the line between heaps and non-heaps?

Many "solutions" have been proposed for this paradox, which to some minds threatened the whole structure of logical deduction, especially in the 19th century when mathematical logic took off. Various treatments by philosophers and logicians are still being published. Current discussions have revolved around questions of ideal languages that contain absolutely no vagueness or imprecision. The treatments produced by philosophers and logicians largely omit references to contemporary situations.

Numericization often violates common sense. We don't seem to be able to escape variations of the sorites paradox. For example, if I had been one year younger, I would not have been subject to mandatory retirement. The question of birthdate comes up regularly when children are assigned an entering class in elementary school, e.g., in one school system, children born after August 31 must wait another year to start. Consider the "appropriate" age for drinking, for driving, or for draft liability. Should it be 16? 18? 21? What determines the distinction between the 1A category and the 4F of a drafted young man? Is the point system banks use for determining whether you are eligible for a mortgage reasonable or discriminatory? From the human point of view, what distinguishes a category 4 hurricane from one of category 5 on the Saffir–Simpson scale? Why is the poverty level set where it is? Do these category distinctions corresponded to some well-marked qualitative change?

In voting, how do we separate candidate A from candidate B and determine who wins an election? Several dozen different voting systems (not hardware) have been proposed and studied. For example there are schemes that employ the majority rule, proportional representation, semi-proportional representation, ranked voting methods, and let's not forget our Electoral College method for determining for presidential elections. Suppose that the Electoral College were replaced by the rule of majority. Suppose, then, that Candidate A received 50,000,000 votes while Candidate B received 50,000,001 votes. (Some recent elections have resulted in very close popular votes.) What to do?

There is a statutory (1941) mathematical formula, known as Huntington-Hill, for determining how many representatives a state may send to Congress after a national census has been taken. Shrieks were heard when New York and Pennsylvania each lost two representatives after Census 2000.

The literature describes numerous criteria that treat voting systems in ways that try to be compatible with commonsense notions of prudence, fairness, and democratic principles. Yet, it turns out that these notions may be mutually incompatible (cf. Arrow's impossibility theorem.)

A numerical index or scheme is an easy and often said to be an objective method for drawing the line somewhere or arriving at a "rational" decision. It may have personal, legal or ethical consequences. And post-determination "fudgings," "tweakings," or creating more "give" in the determinations that jump over boundaries and may create their own ethical problems. The sorites paradox does not go away so easily.

Further Reading

Kenneth Arrow. *Social Choice and Individual Values*. Wiley, 1951.

J. C. Beall, ed. *Liars and Heaps: New Essays on Paradox*. Oxford University Press, 2003.

David M. Farrell. *Electoral Systems: A Comparative Introduction*. St. Martin's Press, 2001.

11

Deductive Mathematics

Mathematics has been called a hypothetico-deductive enterprise. The essential features of mathematical proof can be found by examining Book I of Euclid's *Elements*, one of the most famous books of Western intellectual history. Even though over the years a variety of flaws in the *Elements* have been pointed out and overcome, the spirit of Euclidean deduction pervades all contemporary mathematical developments.

Deductive mathematics, in its formal presentations, at least, begins with definitions and axioms, then proceeds to conclusions arrived at step by step by logical transformations. This process, an ideal, is known as "mathematical proof." Proofs are rarely presented in this precise way, for very good reasons, but the claim is often made that, in principle, it can be done, and this assurance is often sufficient for the mathematical community.

What is an axiom? In an ancient interpretation, an axiom is a self-evident statement, a statement requiring no justification whatever. Although we may ask skeptically, "self-evident to whom?", we may assert that an axiom is (often) an instance of common sense, a commonly accepted notion.

Some examples of axioms:

- Things equal to the same thing are equal to each other.
- If equals be added to equals, the sums are equal.
- Through two distinct points in the plane one and only one straight line may be drawn.

What is a logical transformation? It is often described as a kind of reasoning that no rational, common-sensical person would think to protest.

Some examples of logical transformations:

• If statement A implies statement B and if A is true, then B is true. (Medieval logicians called this the principle of *modus ponens*.) On the other hand, if B is false, then A must be false (*modus tollens*).

• If A implies B, and B implies C, then A implies C. Call this the "domino principle" if you like.

• A implies A or B.

A logical argument or mathematical proof is a concatenation of instances of common sense, and the whole chain of reasoning is part of common sense. My statement about the proof process by itself may not be common sense; it may remind one of the sorites paradox, wherein one argues that since the loss of one hair does not convert a hirsute man into a bald man, then, by iteration, one can conclude that there can be no bald men.

It is also to be noted that in the past century or so, the words "axiom" and "logic" have altered their meaning somewhat. Since formal deduction must begin at some point, an axiom is now considered to be simply such an initial position or gambit, independent of whether any self-evident qualities can be attributed to it.

Consider the famous and notorious "axiom of choice." This asserts that if *C* is a collection of nonempty sets of elements, then we can choose an element from each set in that collection and in this way form another set.

This seems like common sense. If you have a bureau of four drawers, three of which are full of stuff, you can certainly select one thing from each of the three full drawers and throw them into the fourth drawer. The fun—and the doubt—begins when you ask what happens when the bureau has an infinite number of drawers, whatever that means.

The principles of logic have thus been questioned, and there is now an extensive salad bar of alternate systems of logic, among them intuitionistic, many-valued, fuzzy, modal, non-monotonic, and temporal logics. David Mumford envisions a future logic that will embrace probability:

> I will argue that all mathematics arises by abstracting some aspect of our experience and that, alongside the mathematics which arises from objects and their motions in the material world, formal logic arose, in the work of Aristotle, from observing thought itself. However, there can be other ways of abstracting the nature of our thinking process and one of these leads to probability and statistics...
>
> I believe stochastic methods will transform pure and applied mathematics in the beginning of the third millennium. Probability and statistics will come to be viewed as the natural tools to use in mathematical as well as scientific modeling. The intellectual world as a whole will come to view logic as a beautiful elegant idealization but to view statistics as the standard way in which we reason and think.

Each system of logic has its axioms, modes of deductions, procedures, and consequences; we are invited to select whichever system seems sensible, convenient, useful, beautiful or digestible. The varieties of modes of human thought cannot be encapsulated and formalized in a few paragraphs.

Let me give a simple example and a critique of a deductive system. The primitive terms are "person" and "collection."

Definitions. A *committee* is a collection of one or more persons. A person in a committee is a *member* of that committee. Two committees are *equal* if every member of the first is a member of the second, and vice versa. [Note: this is not necessarily true of real-world committees. There may be two committees having different names, purposes, meeting times, etc., but consisting of the same persons.] Two committees having no members in common are called *disjoint* committees.

Axiom 1. Every person is a member of at least one committee.

Axiom 2. For every pair of persons there is one and only one committee of which both are members.

Axiom 3. For every committee there is one and only one disjoint committee.

Theorem. Every person is a member of at least two committees.

Proof:

Statement	Reason
1. Let p be a person.	Hypothesis; naming
2. p is a member of the some committee C.	Axiom 1; naming
3. Let D be the committee that is disjoint from C.	Axiom 3; naming
4. Let r be a member of D.	Definition of *committee*; naming
5. r is not a member of C.	Definition of *disjoint*
6. There is a committee E of which p and r are members.	Axiom 2; naming
7. C and E are not equal.	Definition of *equal*; Steps 5 and 6
8. p is a member both of C and E.	Steps 2 and 6
9. p is a member of at least two committees.	Steps 7 and 8
10. Therefore every person is a member of at least two committees. Q.E.D.	Generalization

Without disputing the conclusion—namely, that it follows from Axioms 1, 2, and 3 that every person is a member of at least two committees—let us examine the claim that the proof material located between the words Proof and Q.E.D. (*quod erat demonstrandum*) constitutes a proof. There is an informal idea that a proof is a sequence of statements written in an unambiguous and strictly formal language that proceeds from the axioms to the conclusion by means of allowed and formalized logical transformations. There is no formal definition of an acceptable proof; a proof that is acceptable to you or to me is one the steps of which we agree constitute a logical sequence.

As we read through the proof, we find that there is one step that is more troublesome than the others. This is Step 7. We pause there, and our minds have to grind a bit before going on. Why are C and E not equal? Spell out the reasons a bit more. They are not equal because r is a member of E by Step 6 but not a member of C by Step 5; therefore, by the definition of the equality of committees, C and E are not equal.

This argument requires that we keep in the forefront of our minds three facts and then verify, mentally, that the situation implies non-equality. This conclusion is deduced from the definition, which speaks only of equality. Thus, in our mind, we have to juggle simultaneously a few more facts: what equality means and how we can proceed to get nonequality out of it.

In order to make clear what is going on, the author of this example attaches to his exposition a symbolic diagram that he says is not really a part of the proof. The picture (remember, it is not part of the proof) supplies the conviction and clarity that are not adequately achieved by the proof. This leaves us with a very peculiar situation: The proof does not convince, and what does convince is not a part of the proof.

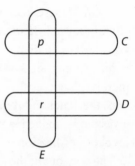

In all human–human interfaces or human–machine interfaces, there is always the problem of verifying that what is asserted to be so is, in fact, so. For example, we assert that we have added two integers properly, or entered such and such data into the computer properly, or the computer asserts that it has carried out such and such process properly. The passage from the assertion to the human acceptance must proceed ultimately by extra-logical criteria.

This problem confronts us constantly. We find in the Reason column of the above proof two mysterious words, "naming" and "generalization," used in the proof. Now, if there is nothing worth discussing about these ideas in their application to the proof, why did the author bother mentioning them? Both are, in fact, difficult concepts, and philosophers have dedicated whole books to them. If they are important in the present context, how do we verify that the naming process or the generalization process has been carried out properly?

Look at generalization. In Step 1, a typical person is selected and named. Since it is a typical person, it is not specified which person it is. The idea is that if one reasons about a typical person, and uses only the characteristics that that person shares with all other persons, then one's deductions will apply to all persons (Step 10). Should it not be verified, then, as part of the proof, that only those characteristics have been used? What are the formal criteria for so doing? By raising such questions, one can force the proof into deeper and deeper levels of justification. What stands in the Reason column now, the single word "generalization," is pure rhetoric.

A rather different point is this. Suppose we have set up certain abstract axioms. How do we know that there exists a system that satisfies those axioms? If there is no such system, then we are not really talking about anything at all. If there is such a system, its existence might be made known to us by display: "Such and such, with such and such definitions, is an instance of a system that fulfills Axioms 1–3." Would we then merely glance at this statement and agree with a nod of the head, or does the statement require formal verification that a purported model of a system is, indeed, a model? Again, we have been driven into a deeper level of verification.

The way out of these difficulties is to give up the needless and useless goal of total rigor or complete formalization. Instead, we recognize that a mathematical argument is addressed to a human audience, which possesses a background knowledge enabling it to understand the intentions of the speaker or author. In stating that a mathematical argument is not mechanical or formal, we have also stated implicitly what it is—namely, a human interchange based on shared meanings, not all of which are verbal or formulaic. That is the common sense of the situation.

Further Reading

Philip J. Davis and Reuben Hersh. "Rhetoric and Mathematics." In *The Rhetoric of the Human Sciences,* John S. Nelson, et al., eds. University of Wisconsin Press, 1987.

Thomas L. Heath. *The Thirteen Books of Euclid's Elements.* Dover (Reprint), 1956.

David Mumford. "The Dawning of the Age of Stochasticity." In *Mathematics: Frontiers and Perspectives*, pp. 197–218. American Mathematical Society, 2000.

Richard Trudeau. *The Non-Euclidean Revolution.* Birkhäuser, 1987. Courtesy of the author, the example of a deductive system just given has been reproduced with a few changes from this book.

Robert S. Wolf. *A Tour through Mathematical Logic.* Mathematical Association of America, 2005.

Mathematics Brings Forth Entities Whose Existence Is Counterintuitive

The power of square roots is an intangible power: their rigid laws have survived for centuries.
—Umberto Eco

Science can be, often should be, a nuisance to the established order...
—Joshua Lederberg

Mathematics is true facts about imaginary objects.
—Reuben Hersh

There are mathematical entities, objects, or processes that do not accord with intuition or common sense, and mathematics keeps growing through the creation of new ones. Explanations of the existence of such things are part of what philosophers call mathematical ontology. These are objects that cannot exist within certain standardized mathematical theories, but when the mathematical environment is enlarged, they can exist and thrive. Their existence may be justified on the basis of unification and extension of ideas and operations, and their continued existence derives from utility within either pure or applied mathematics. Thus sense has been granted them. The opposite has also occurred: An idea was accepted and later, when examined, sense was denied. Mathematics, like the law, can change its mind.

Mathematical Objects

The very words "zero," "negative number," "imaginary number," and "irrational number" imply doubts. Existential angst regarding complex imaginary numbers and infinitesimals lingered on for years, and even penetrated the textbooks of my college days.

Zero (0). Common sense complains: How can nothing be something (an entity)? Numbers arise from counting objects. How can no objects be counted? There were no zeros among the classical Greeks or Romans, but the Indians had it around the fifth century. In set theory, we learn about the *null set*, the set that has nothing in it. How can there be a set that has nothing in it? Can there be two different null sets, or zeros?

The concept of zero proves useful in arithmetic and in many other places. Zero is self-reproducing under addition: $0 + 0 = 0$, and any number multiplied by zero yields zero. Restrictions are placed so as to promote meaning, to avoid paradox and restore common sense. As an example, since $6 \times 0 = 5 \times 0$, then dividing both sides of the equation by 0 yields $6 = 5$, a contradiction. Thus, never, ever divide by zero. Yet, differential calculus can be based on a kind of division by zero, properly construed. Another caution of some years ago: Never, ever work with a divergent series, yet, Euler did it fruitfully.

One (1). Common sense complains: How can 1 be a number? A number should express numerosity or multiplicity. There is a long historic and philosophical record of 1 not being considered a number. It was the basic unit from which other numbers were constructed but not itself a number. The concept of 1 is useful; it is self-reproducing under multiplication: $1 \times 1 = 1$.

Fractions (1/2, 2.85, etc.). Common sense complains: How can $\frac{1}{2} = \frac{2}{4}$ when a half a pie is clearly not the same as two pieces of a pie cut into quarters? How can the average number of children in an American family be 2.85? To the ancients, a number was a whole number, i.e., an integer. A fraction was not a quantity in its own right. Carl B. Boyer and Uta C. Merzbach point out, "The Egyptians regarded the general proper rational fraction of the form $\frac{n}{m}$ not as an elementary 'thing' but as part of an incomplete process [that related it to unit fractions of the form $\frac{1}{n}$]." The division of all numbers by all numbers (with the one exception: no division by zero) came in gradually. We teach children the arithmetic of fractions, and they learn it momentarily. But how often does a sum such as $(\frac{2}{7}) + (\frac{347}{581})$ arise in practice?

Irrational numbers ($\sqrt{2}$, $\sqrt{6.7}$, etc.). In 550 BC, the concept of $\sqrt{2}$ raised eyebrows. It existed as the length of the diagonal of the unit square. It didn't exist because, as the Pythagoreans found, it could not be expressed in terms of the kind of numbers that were then conceivable. This is the crisis or the paradox of Pythagoras.

Forced into existence by geometry, √2 exists as an infinite sequence of integers:

$$\sqrt{2} = 1.41421356237309504880168872420969807856967187537694807317667973799 0...$$

Negative numbers (e.g., –7). Common sense asks "How can less than nothing be something?" After all, there can't be minus three gallons of gas in my gas tank. (Consider the common expression "driving on empty.") How can there be minus three dollars in my bank account? But the bank occasionally allows this and makes a profit on doing so. Negative numbers make all subtractions possible: $8 - 10 = -2$. Other operations with negative numbers, such as $-2 \times (-3) = +6$, raise their own questions. How can something positive emerge from negatives? Beginners are often confused by this. Notice, though, that in ordinary language a double negative is often a positive. E.g., she not infrequently goes out for a walk.

Imaginary or complex numbers (e.g., √–1, 6 – 2√–1). Common sense asks "How can the laws of the multiplication of positive and negative numbers be violated?" The product of a positive number and itself is positive. The product of a negative number and itself is also positive. But the product of the square root of a negative number and itself would be negative. Contradiction.

The famous British mathematician and logician Augustus De Morgan (1806–1871) was plagued by ontological and epistemological doubts about formulas involving complex numbers. Here I quote a somewhat tongue-in-cheek, I think, passage:

> Imagine a person with a gift of ridicule. [He would remark] first that a negative quantity has no logarithm; secondly that a negative quantity has no square root; thirdly that the first non-existent is to the second as the circumference of a circle is to the diameter.

Here De Morgan is reformulating the famous Euler identity $e^{\pi i} = -1$.

Infinitesimals (e.g., *dx*). Complaint: How can there be a positive number that is simultaneously less than 1/2, less than 1/3, less than 1/4,... less than, in fact, all such fractions $1/n$, and still not equal to zero? This would seem to contradict common sense. It bothered Bishop George Berkeley, but it didn't seem to bother Leibniz who used them as the basis of his differential calculus. Utility can often dissipate existential angst, but the angst may still linger. The 19th century mathematicians got around the strained logic of infinitesimals through the introduction of the limit concept.

An anecdote: William Fogg Osgood, distinguished mathematician and professor at Harvard, author of undergraduate texts and of the very influential *Lehrbuch der Funktionentheorie,* thought that the formal discussion of limits was too difficult for Harvard freshman to conceptualize, so he used the older notion of infinitesimals in his book on elementary calculus. A later generation of instructors, required to use Osgood's text, advised students to rip out the section on infinitesimals.

There is now a complete theory of infinitesimals that overcame Berkeley's existential angst. They are part of what is called the hyperreal or non-standard number system. Don't try to locate infinitesimals in any computational software, however; you will not find them there. Non-standard calculus texts have been written but are not popular and are probably unnecessary.

Quaternions (William Rowan Hamilton, 1843). Complaint: How is it possible that *a* times *b* is not equal to *b* times *a?*

Comment: If *a* and *b* are conceptualized as rotations in space, and multiplication is conceptualized as one rotation followed by another, then usually $a \times b \neq b \times a$. Quaternions are the first example of what are now known as non-commutative algebras.

Infinity (∞, $\aleph_0$, etc.). Common sense denies the infinite, and there are numerous varieties of the denial. What sense does it make to write down a number so great that it exceeds any quantity that arises in physical theory? How is it possible to count endlessly? All things must end. How can there be a greatest number when you can add one to it and get something still greater? Infinity is often thought of as the greatest number and is represented by the symbol ∞. In point of fact, there are numerous conceptualizations and uses of infinity. Making sense out of the product $\infty \times 0$ can, in some sense, be considered the basis of the integral calculus. The notion of infinity has also been allied to theological concepts, which accords it a certain credibility.

Galileo noticed that there are as many square integers 1, 4, 9, 16, ... as there are integers 1, 2, 3, 4, ... Common sense asks, How is this possible? After all, the set 1, 2, 3, 4, ... contains many more integers than the set 1, 4, 9, 16, ... This claim depends upon what you mean by the phrase "as many." Then, lo and behold, along came Georg Cantor, who defined the phrase in such a way that it led to this counterintuitive claim and also to a major field of mathematics (*Mengenlehre*, or set theory). The multiplicity of the set 1, 2, 3, 4, ... and of any set that can be put into a one-to-one correspondence with it s denoted by $\aleph_0$, and with it begins an arithmetic of the infinite.

Paradoxes of the infinite. The status of arguments that go from the finite to the infinite is tricky. Here is an instance of such an argument that is defective. The finite Taylor expansions of the exponential function e^x, $1 + x + x^2/2! + x^3/3! + ... + x^n/n!$, are polynomials of degree n and consequently have n roots. One might expect, then, that since $e^x = 1 + x + x^2/2! + x^3/3! + ... + x^n/n! + ...$ (which is, so to speak, a polynomial of infinite degree), it would have an infinite number of roots. But no, in fact, e^x has no roots whatsoever. On the other hand, the argument works for the polynomials $1 - x^2/2! + x^4/4! - x^6/6! + ... \pm x^{2n}/(2n)!$, which approximate the periodic function $\cos x$ and have two roots within each period.

Common sense leads us to wonder about the many paradoxes that are embodied in the infinite and is often ready to throw the concept out the window. Here's an old mathematical chestnut.

Set $\qquad\qquad S = 1 - 1 + 1 - 1 + 1 - 1 + ...$

Then $\quad S = (1 - 1) + (1 - 1) + (1 - 1) + ... = 0 + 0 + 0 + ... = 0.$

On the other hand,

$\qquad\qquad S = 1 + (-1 + 1) + (-1 + 1) + ... = 1 + 0 + 0 + ... = 1.$

Hence $0 = 1$, a contradiction. Apparently one should never, ever work with a divergent series. Yet Euler did it fruitfully! The contradiction is cleared up by the theory of convergent and divergent infinite series, which become important chapters of mathematical analysis.

The mathematical line. Common sense asks: How can there be a line or a curve with zero thickness that is (occasionally) of infinite length? Mathematics ignores common sense by idealizing the physical world.

Non-Euclidean geometry. Euclidean geometry says that through a given point P that is not on a given straight line L one and only one straight line exists that is parallel to L, Non-Euclidean geometry (there are several types) says that through a given point P not on a given line L you can draw either an infinite number of parallels to L or no parallels. How's that? Common sense reinforced by Euclid and Kant said: this is nonsense. The allowability of the axioms of non-Euclidean geometry was disputed furiously. In mid-Victorian England this argument raised almost as much heat as Darwin's theory of evolution.

Space filling curves (Giuseppe Peano, 1890). Common sense asks: How can a two-dimensional space, which has breadth and width, be totally filled by a curve that lacks thickness?

Approximation to a Peano curve

Visual images break down here. A Peano curve is defined by means of formulas, but the above graphical approximation provides a clue as to how this may be possible.

Continuous non-differentiable curves. By "the direction of a curve at a point" is meant the direction of the line tangent to the curve at that point. Claim: There are continuous curves that have no direction at any point (non-differentiable curves). If you could travel along such a curve you would never know in which direction you were going!

Visuals fail in this situation. A creature such as a non-differentiable curve must be defined by formula. But the above picture of the approximation to the Peano curve suggests how this may come about. Take a smooth curve. Put a lot of corners in it. Put more corners where it still remains smooth. Carry out this operation over and over again. Eventually (whatever that means) you arrive at a non-differentiable curve.

The Dirac function $\delta(x)$. Complaint: How can a "curve" (or function) that is zero everywhere except at one point, have a positive area underneath it?

Visuals are only suggestive here.

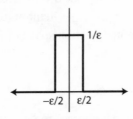

The Dirac function, often called the impulse function, can be defined by many formulas such as the limit as $\varepsilon \rightarrow 0$ of the function depicted above, or as

$$\delta(x) = \lim_{w \to \infty} \frac{\sin(2\pi w x)}{\pi x}, \, w > 0$$

which displays it as a limiting wave packet. The legitimacy of these formulas in the sense of axiomatic mathematics is guaranteed by such generalized function theories as "distribution theory."

Higher-dimensional spaces. Complaint: When we move around, we experience only length, breadth, and thickness. Some would add "and duration." Though interesting representations of four-dimensional objects were made quite a long time ago and more recently through computer animation, we do not experience four or greater dimensions via naïve visualization.

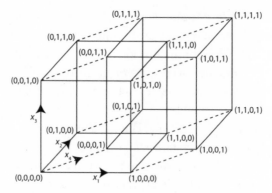

A four-dimensional cube

Since the work of Descartes, two-dimensional space has been represented by a pair of numbers (x, y) conceived of as an object in its own right. Three-dimensional space is represented by a triple of numbers (x, y, z). It is easy to say, then, that n-dimensional space is simply an n-tuple of numbers $(x_1, x_2, x_3,..., x_n)$. Quantum theorists have spoken of particles embedded in a configuration space of $3^{10^{17}}$ dimensions, which surely boggles the mind.

Mathematical Processes

Induction. Warnings have been given to encourage procedural or heuristic caution in "infinite situations"; one cannot (and therefore should not) draw any conclusions from a finite number of cases. As a methodological precept, this is palpable nonsense. Fruitful conclusions are drawn every day of the week from a finite number of cases. This is what the inductive method is all about. This is what the statistical method is all about. But these methods must be accompanied by a dose of common sense skepticism, for people are fond of jumping to conclusions.

Long proofs. Should we stomach long (but traditionally fashioned) proofs, yes or no? A conversation with the late Daniel Gorenstein revealed his life goal of reducing the 15,000-page proof of the classification of simple finite groups down to a mere 5000 pages. What one person could read through and check the details of such material? Such a person would be a very rare human. What would be the credibility status of such a proof? Problematic. Would eyebrows be lowered a bit after the reduction to 5000 pages? Not one whit.

Continuous vs. discrete. The continuous–discrete split is often rancorous. In philosophical discussions it goes under the tag line *natura non facit saltum* (nature doesn't make jumps). The argument rears its head, for example, in biology and in continuum mechanics. In these fields there are those who want to pass from the atoms (at the micro level) to properties of materials (at the macro level) by discrete means. Others say this is not possible. The split has entered the discussions of mathematical curricula for the 21st century: Is continuous mathematics (i.e., the calculus and all its developments) of decreasing relevance in applications? Common sense here seems to hide in the bushes.

Further Reading

Carl Boyer and Uta C. Merzbach. *A History of Mathematics,* 2nd. ed. Wiley, 1989.

Philip J. Davis. "The Aesthetic Impulse in Science and Mathematics." *SIAM News*, Vol. 38, No. 8, Oct. 1999.

Philip J. Davis. "The Unbearable Objectivity of the Positive Integers." *SIAM News*, Vol. 32, No. 5, June 1999.

Philip J. Davis. "Paraconsistent Thoughts about Consistency." *Mathematical Intelligencer*, Vol. 24, No. 4, 2002.

Philip J. Davis. "Mathematics and Common Sense: Cooperation or Conflict." In *Proceedings of the 47th CIEAEM Meeting: Mathematics (Education) and Common Sense*, Freie Universität Berlin, 1996.

Philip J. Davis. "Visual Geometry, Computer Graphics and Theorems of Perceived Type." *Proc. of Symposia in Applied Mathematics*, Vol. 20, American Mathematical Society, 1974.

Philip J. Davis and David Park, eds. *No Way: The Nature of the Impossible*. W. H. Freeman, 1987.

Robyn Dawes. *Rational Choice in an Uncertain World*. Harcourt Brace Jovanovich, 1988.

Robert Kaplan. *The Nothing That Is: A Natural History of Zero*. Oxford University Press, 1999.

Graham Oppy. *Philosophical Perspectives on Infinity*. Cambridge University Press, 2006.

Joan L. Richards. *Mathematical Visions: The Pursuit of Geometry in Victorian England*. Academic Press, 1988.

Constance Reid. *From Zero to Infinity: What Makes Numbers Interesting,* 5th ed. A K Peters, 2006.

Brian Rotman. *Signifying Nothing: The Semiotics of Zero.* St. Martin's Press, 1987.

Brian Rotman. *Ad infinitum—The Ghost in Turing's Machine: Taking God Out of Mathematics and Putting the Body Back In.* Stanford University Press, 1993.

Robert S. Wolf. *A Tour Through Mathematical Logic.* Mathematical Association of America, 2005.

For more information about √2, see the following website:
http://antwrp.gsfc.nasa.gov/htmltest/gifcity/sqrt2.1mil

"In Principle We Can...": Mathematical Existentialism

The phrase "in principle we can..." occurs frequently in mathematical and scientific expositions. There are many principles of thought or action that lie beneath the standard axiomatizations or formalizations. These principles are never mentioned explicitly in texts, but are learned and accepted as part of the craft. What does the expression mean? It generally appears in the positive form: "In principle we can...", but a negative version also exists: "Even in principle we cannot..."

"In principle we can count the number of acorns that fell to the ground in New England last November." An ecologist might be able to make a reasonable estimate, but it would not involve counting: 1, 2, 3, ... up to the last nut.

"In principle we can tell how many people are in Times Square on New Year's Eve." One can estimate as follows: Take a picture. Divide the photo into one-inch or half-inch squares and count the number of people in one square (or take an average of the number of people in five squares). Multiply by the number of squares.

"Using the pigeonhole principle, we can prove that there exist at least two people in London that have the same number of hairs on their head." The number of individual hairs on a person's head is said not to exceed 200,000. But can we locate the two people?

The pigeonhole principle (also called the Dirichlet box principle) is an example of a counting argument that can be applied to many "more serious" mathematical problems. A generalized version of the principle states that, if m discrete objects are placed in k boxes, then one box contains at least $\lceil m/k \rceil$ objects, where $\lceil x \rceil$ denotes the ceiling function of x, i.e., the least integer greater than or equal to x.

Pierre-Simon Laplace's "demon" is an excellent example of an "in principle statement." Here are Laplace's exact words:

> We may regard the present state of the universe as the effect of its past and the cause of its future. An intellect which at any given moment knew all of the forces that animate nature and the mutual positions of the beings that compose it, if this intellect were vast enough to submit the data to analysis, could condense into a single formula the movement of the greatest bodies of the universe and that of the lightest atom; for such an intellect nothing could be uncertain and the future just like the past would be present before its eyes.

This means that, in principle, if we knew the positions and velocities of all the particles in the universe, then using Newton's Laws of motion, we could compute the future state of the universe at all later times. Not a good computational option!

"In principle, we can list all the fractions one after another in a sequence." We can certainly give a recipe for doing it, but our lives being finite, and the fractions being infinite in number, we can only imagine such a list.

"In principle we could examine the approximately 300,000,000 cases of whatever, one by one, and then decide what the situation is." But it might be common sense to make the decision another way.

"In principle we can take all the U.S. social security numbers in existence today and concatenate them into one huge integer." We could probably do this with today's computers, but to what purpose?

"In principle we can derive any mathematical result, logical step by logical step, from the axioms." But no one ever has written down all the consecutive steps for the vast corpus of mathematical theorems, and no one will ever do it. There is no need to do it.

So what are the various interpretations of "in principle"? When philosophers use it, it typically means "it is possible that," but whether the possibility is logical or more substantive is usually not specified. Sometimes the expression means the possibility doesn't contradict known laws of physics, mathematics, etc.

The phrase is often omitted. Consider, for example, the famous Axiom of Choice (AC), often formulated roughly as follows.

Given any set S of mutually disjoint nonempty sets, then there is a set C containing one member from each element of S.

C can thus be thought of as containing a chosen representative from each set in S, hence the name Axiom of Choice. Would it be any weaker, or more comprehensible, if instead of "there is a set," we said "in principle we can form a set"?

Discussions of the phrase "in principle" drive us to clarify the meanings of such words as "find," "exist," "construct," "choose," "list," "build," "let," and "consider" and to distinguish these verbs as designating a purely mental act, differing from a physical act carried out in the real world. There is a fundamental gap between existence as logical possibility and existence as construction. If we insist on a physical interpretation of construction, then the phrase "in principle" can become a scam, a swindle. If we accept this interpretation, then mathematics is a world of mental fictions, harboring statements of which some are commonsensical, some nonsensical, and some neither one nor the other.

Further Reading

Reuben Hersh. "Mathematical Practice as a Scientific Problem." In *Current Issues in the Philosophy of Mathematics From the Viewpoint of Mathematicians and Teachers of Mathematics*, Bonnie Gold and Roger Simons, eds. Mathematical Association of America, 2006.

Brian Rotman. *Mathematics as Sign: Writing, Imagining, Counting.* Stanford University Press, 2000.

14

Mathematical Proof and Its Discontents

What Is a Proper Mode of Mathematical Argumentation?

Insofar as mathematics is validated by the agreement of the community of mathematicians and those who employ it, this common agreement constitutes a kind of common sense. At any time and place, therefore, within mathematics, its theory, applications, and development, there are operative common sense, intuition, induction, argumentation, deduction, and numerous other principles of growth and regulation. Aside from mathematical deduction in the sense of Euclid, there are other modes of deduction and argumentation validated by other intellectual communities that also constitute expressions of common sense and may be in conflict.

Consider some historical and simple instances where mathematics is (or was) in conflict with common sense. Nicholas of Cusa (Cusanus, c. 1450) was a cardinal of the church and possessor of a fine mind. Cusanus thought that the circle could be squared. An incomplete description of the meaning of "squaring a circle" is as follows: Using a finite number of ruler and compass constructions of a certain type, construct a square whose area is equal to that of a specific circle. Since classical antiquity, when the playwright Aristophanes ridiculed the "circle-squarers" in one of his comedies, the common sense of the matter was that this was impossible.

How did Cusanus arrive at this conclusion? He believed in the philosophical doctrine of the concordance of contraries. This led him to think that maxima and minima are related, and hence that the circle (a polygon with the greatest number of sides) must have the same properties as the triangle (the polygon with the least number of sides). Since a triangle can be squared, it follows that the same must be true of a circle.

Though Cusanus opposed certain scholastic dogmatisms, his sort of argument was, in his intellectual circle, reasonable and commonsensical. After all, in his day, and even now, it has been productive to think of a circle as a regular polygon with an infinite number of sides. But the question of whether one can carry over to an infinite case a situation that prevails in all finite cases is a very delicate one, and the answer may be either "yes" or "no."

In Cusanus' day, the question of the squaring of the circle had not yet been settled in the sense of axiomatic argumentation of the Euclidean type, and it would hardly have been possible then to come forward with the algebraic and analytic ideas through which the impossibility was ultimately proved in the late 19th century.

A second example of a kind of spurious reasoning applied to mathematics. The philosophers George Berkeley and David Hume here (the latter in his *Treatise on Human Nature, Part 2, Book 1*) both asserted vigorously that a straight line cannot have an infinity of points along it. But ask yourself, how can this be the case when, given two distinct points A and B on a line, there is always a midpoint between them that is distinct from either A or B?

Or is there? Were these two philosophers perhaps thinking of the straight line as a physical object? Whatever reasons lay behind their conclusion, a modern critic might say that they had confused an idealized mathematical line with a real world line drawn with a ruler and pencil. The graphite trace on a sheet of paper is subject to physical analysis, and an assertion of finiteness might be made based on theories of particle physics. An idealized line exits only in the world of ideas, and conclusions about it are reached by the interplay of axiomatic ideas.

In contrast to Euclidean geometry, finite geometries having a finite number of points have been developed in the 20th century. Gino Fano (1871–1952) discovered that the axioms of Euclidean geometry did not imply an infinite number of points and took it from there. Finite geometries have found applications in, e.g., the theory of sampling.

A take-home problem: How it is that intelligent people with a common fund of information can reach different conclusions and cannot be persuaded to change them?

Difficulties with Verification

The famous Kepler conjecture asserts that the optimum arrangement for packing equal spheres in terms of average density is that of the "fruit sellers" arrangement. (Think of the pile of grapefruit in the fruit section of your market.) This conjecture, which seems eminently sensible, turned out to be extremely difficult to prove mathematically.

Finally, in 1998, Thomas Hales gave a proof of Kepler's conjecture: The proof is over 250 pages of text and 3 gigabytes of computer code and output. The *Annals of Mathematics* solicited the paper for publication in 1998 and hosted a conference in January 1999 that was devoted to understanding the proof. A panel of 12 referees was assigned to the task of verifying the correctness of the proof. After four full years, Gabor Fejes Tóth, the chief referee, returned a report stating that he was 99% certain of the correctness of the proof, but that the team had been unable to certify the proof completely.

Robert MacPherson, editor of the *Annals*, wrote a report that stated

> The news from the referees is bad, from my perspective. They have not been able to certify the correctness of the proof, and will not be able to certify it in the future, because they have run out of energy to devote to the problem. This is not what I had hoped for.

MacPherson continues with comments to Hales that may well mark a profound change in the way that mathematics is done.

> The referees put a level of energy into this that is, in my experience, unprecedented. They ran a seminar on it for a long time. A number of people were involved, and they worked hard. They checked many local statements in the proof, and each time they found that what you claimed was in fact correct. Some of these local checks were highly non-obvious at first and required weeks to see that they worked out. The fact that some of these worked out is the basis for the 99% statement of Fejes Tóth that you cite. One can speculate whether their process would have converged to a definite answer had they had a more clear manuscript from the beginning, but this doesn't matter now. Fejes Tóth thinks that this situation

will occur more and more often in mathematics. He says it is similar to the situation in experimental science—other scientists acting as referees can't certify the correctness of an experiment, they can only subject the paper to consistency checks. He thinks that the mathematical community will have to get used to this state of affairs.

You may ask whether this degree of certification is enough checking for a mathematical paper, and whether it's not in fact comparable to the level of checking for most mathematical papers. Both the referees and the editors think that it's not enough for complete certification as correct, for two reasons. First, rigor and certainty was what this particular problem was about in the first place. Second, there are not so many general principles and theoretical consistencies as there were, say, in the proof of Fermat, so as to make you convinced that even if there is a small error, it won't affect the basic structure of the proof.

The *Annals* is confident enough in the correctness that they will publish a paper on the proof, but a short note from the editors will state that they have been unable to certify it completely.

Further Reading

F. Karteszi. *Introduction to Finite Geometries*. North-Holland, 1976.

For more on the Kepler conjecture, see the following websites:
http://www.math.pitt.edu/~thales/flyspeck/
http://mathworld.wolfram.com/KeplerConjecture.html

The Logic of Mathematics Can Spawn Monsters

Itaque elegans et mirabile effugium reperit in illo Analyseos miraculo, idealis mundi monstro, pene inter Ens et non-Ens Amphibio, quod radicem imaginariam appellamus. [The imaginary numbers are a delicate and wonderful refuge of the human spirit, a creature that lives in an ideal world, almost an amphibian between being and not being.]
—Gottfried v. Leibniz, quoted in J. Tropfke, *Geschichte der Elementaren Mathematik*, 1980, p. 153

[One of the professors in the lunatic asylum] explained to me that inside the globe was another globe much bigger than the outer one.
—Jaroslav Hašek, *The Good Soldier Švejk*

Logic is the art of going wrong with confidence.
—W. H. Auden, who ascribed this quip to Anonymous

A monster is unusual, unfamiliar, totally out of the range of everyday or common-sense experience, and often frightening. Monsters play a considerable role in our imaginative lives: the manticore of the mediaeval bestiaries, the giant in *Jack and the Beanstalk,* Cookie Monster of Sesame Street. There are monstrous individuals, things, acts, thoughts, places. There are even mathematical monsters.

Over the years, logic, as plied in mathematics, has spawned monsters, and hence, in the view of some, logic has taken a beating. In the view of others, the workings of logic have enriched the world with imaginative possibilities. What is thought to be logically impossible has been identified with the ridiculous, incredible, nonsensical, monstrous. Later, what was once monstrous has often turned out to be logically possible, consistent, and accepted; gradually the cachet of the monstrous has abated, and

The Arch in St. Louis
(Courtesy: Kevin Jackson-Mead)

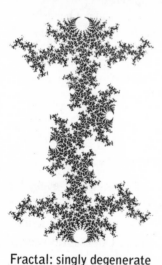

Fractal: singly degenerate Julia set
(Courtesy: David Mumford, Caroline Series, and David Wright)

the monster has been tamed and has become part of work-a-day mathematics.

Some of the most recent instances of mathematical monsters are the images of fractals. They are certainly weird visually, far different from the usual beautiful mathematical curves, conic sections, spirals of Archimedes and Bernoulli, epic- and hypocycloids, etc. The inverted mathematical catenary, built into the Gateway Arch in St. Louis, is breathtaking in its beauty and sweep. In contrast, fractal images often appear rough, fragmented, vermicular, other-worldly, and monstrous. And yet, these images and their theories appear to simulate nature in the shapes of clouds, coastlines, and many such shapes have beauty in them.

One of the mathematical monsters spawned by logic is a theorem that goes by the name of the Banach-Tarski-Hausdorff paradox (1924 and before). To state the paradox one way, you can take a solid sphere, break it into a finite number of pieces, and reassemble the pieces in such a way to form two solid spheres each of volume equal to the original. (It can be done with five pieces.) Common sense says: how ridiculous!

To formulate the paradox more carefully and a bit differently: Define two finite sets of regions $A_1, A_2, ..., A_k$ and $B_1, B_2, ..., B_k$ such that

- For each $i = 1, 2, ..., k$, A_i is congruent to B_i in the Euclidean sense.
- The A_i's are pairwise disjoint, as is the case for the B_i's.

- The union of the A_i's is a small sphere, while the union of the B_i's is a big sphere.

How is this possible? It seems like utter nonsense. If possible, we could take a costly gold sphere, invoke the construction of the paradox, and get two copies to double the value. If we kept multiplying spheres, we would soon become wealthier than Midas.

So what is going on here mathematically? In order to understand, we must first get rid of some dearly held ideas.

- A mathematical sphere is not a physical sphere, nor is a mathematical piece a physical piece.
- While physical "pieces" have volumes, mathematical pieces (i.e., sets) can be such that their volumes cannot be defined. The pieces of the B-T-H paradox are non-measurable sets.
- How is the breaking apart done? An orange or a block of wood can be taken apart by a knife or a saw, but the mathematical sphere of the paradox is taken apart by invoking the controversial axiom of choice.
- How is the reassembling done? Certainly not along edges: There are no edges to the pieces.

These considerations demonstrate vividly the extent to which so-called logical thought can be at odds with common, everyday experience. But an endlessly alluring aspect of mathematics is that its thorniest paradoxes have a way of blooming into beautiful theories.

Further Reading

Solomon Feferman. "Does Mathematics Need New Axioms?" *American Mathematical Monthly*, Vol. 106, No. 2, 1999.

Solomon Feferman. "Mathematical Intuition vs. Mathematical Monsters." *Synthèse*, Vol. 125, 2000.

Robert M. French. "The Banach-Tarski Theorem." *Mathematical Intelligencer*, Vol. 10, 1988.

Thomas J. Jech. *The Axiom of Choice*. North-Holland, 1973.

Stan Wagon. *The Banach-Tarski Paradox*. Cambridge University Press, 1985.

Leonard M. Wapner. *The Pea and the Sun: A Mathematical Paradox*. A K Peters, 2005.

16

Rules and Their Exceptions

Common sense tells you to neglect the exceptional and live within the known world. But art and science are for a moment one in the injunction, even the commandment, to look first, only, always, at the exception, at what doesn't fit: because, one says, it will turn into the universal while you look; because, says the other, it will show you the way to a universal not yet known.

— Howard Nemerov

We make rules, and having made them, we almost immediately find we have to plug them up with exceptions or qualifications. And very often the exceptions have exceptions. Every rule appears to have an exception, but there are exceptions to this rule!

Social. "Garden Party, 4:00 PM on June 25th. Rain date (i.e., the exception and the default option, the "else") June 26th."

Law. It is against the law to attack a person, except in self defense, and the Fourth Amendment prohibition of searches and seizures has many exceptions.

Pharmaceutical. "You may take this medication with all fruit juices except grapefruit juice."

Programmatic. Consider the conjunctive coding instructions that build in exceptions characterized by the following expressions: which of, which if, which unless, else, default.

Dining halls. As an undergraduate, the dining hall rule in my college was that you could order one of two entrees for supper. (We were served by student waiters and professional waitresses, not waitpersons.) If you liked neither, then, as a default option, you could order a can of sardines on a bed of lettuce.

Spelling. "I before E, except after C." There are so many exceptions to this rule (heifer, neither, science, foreign, weird, as starters) that super-rules have been formulated. These are so complicated that they are practically worthless. No wonder foreigners tear out their hair when learning English spelling.

Exceptions in Mathematics

Part of the problem here is linguistic, and part is our insufficient experience with mathematical objects. On the whole, mathematicians abhor exceptions, and once exceptions are discovered they are often spirited away by redefinitions. Although the process may seem a scam, it can suggest stimulating new theories.

Defining a triangle, one may say "Any three points in the plane determine a triangle, and the three points are its vertices." Exception: The three points must not lie on a straight line. One often says "In general, three points determine a triangle." The phrase "in general" sweeps aside all exceptions in a vague way.

A second way to define a triangle: Any three non-collinear points in the plane determine a triangle. (Here the rule "swallows" the exception.)

A third definition: Extending the notion of a triangle to include the "null" triangles of zero area whose vertices lie on a straight line. Any three points in the plane determine a triangle.

Remark. In the case of three collinear points, the formula for the area of a triangle in terms of the coordinates of the points, as well as the formula for the area in terms of its sides, will both give zero. This fits in well with our intuitive or visual notion of area.

Theorem. Any two circles in the plane coincide in exactly two points.

Exceptions. If the centers of the circle are too remote from each other, the circles don't coincide at all. If one circle lies inside an outer circle, they may not coincide at all. If the two circles are tangent to each other, they coincide in either one or in all their points. All these come from simple observation. But, as remarked, mathematicians don't like exceptions; they look around for unifying principles. The point at which two circles are tangent is called a double point, so this exception is swallowed by the rule.

Matters don't, of course, remain so simple. When complex projective geometry appeared on the scene with Jean-Victor Poncelet (1788–1867), it turned out that every circle passes through two so-called imaginary circular points at infinity, and hence two circles always coincide in at least two points. I leave to my readers the joy of formulating a general rule by consulting Bezout's Theorem and then describing the possible coincidences of three or more circles.

I list several more theorems, not to focus on their content but to show how exceptions and no exceptions get threaded linguistically.

Theorem. With the exceptions of $n = 4, 6, 8, 12$, and 20, there are no regular platonic solids with n faces. (The theorem is not usually stated in this negative way.)

Theorem. Every square matrix, with many exceptions, may be diagonalized. Every square matrix, with no exceptions, may be put into Jordan form.

Theorem. Every matrix has a Moore-Penrose generalized inverse. The exception of non-singularity and matrix shape has been totally eliminated by extending the notion of inverse.

Imre Lakatos used the word "monster" to describe a counterexample or exception that demolished a conjecture. A monster was a discovery that gave rise to a modification of old theoretical statements and often to new concepts. The conjecture is maintained by the process of "monster-adjustment" or "exception-barring" that restrict the domain of the conjecture. Context extension also maintains the conjecture.

Further Reading

Claire Oakes Finkelstein. "When the Rule Swallows the Exception." *Quinnipiac Law Review Association*, Vol. 19, No. 3, 2000.

Imre Lakatos. *Proofs and Refutations*. Cambridge University Press, 1976.

Howard Nemerov. *Figures of Thought: Speculations on the Meaning of Poetry and Other Essays*. David Godine, 1978.

If Mathematics Says "No" Does It Really Mean It?

Experience is never limited and never complete.
—Henry James

To try to do something which is inherently impossible is always a corrupting enterprise.
—political philosopher Michael Oakeshott
(Except in mathematics where doing the impossible can be a creative act.)

Mathematical Statements of Impossibility

When mathematics says no, is that the final word? When all the impermanencies of the world are considered, when one thinks of vast empires that have fallen, of religious belief and customs consigned to the ashheaps of time, of facts and systems of science patched up as the result of body blows received from pummeling nature, when one sees day-to-day arrangements of life changing rapidly before our eyes, in what quarter are we to find a yearned-for permanence?

For a long time, one answer has been that permanence and security can be found in the realm of mathematics. The proven statements of mathematics have been considered true and indubitable, and it has been believed that they are universal, that their truth is independent of time and of national (or even intergalactic) origin. Some mathematicians think of them as the common sense underlay of mathematics. Since these commonly held views are by no means self-evident, they have naturally been the subject of much discussion. Over the years such discussions constituted a significant part of what is called the philosophy of mathematics. In my opinion (and that of many observers of the mathematical scene), these views are naïve and lead to an inadequate picture of mathematical activity.

In this essay, I shall explore these views from a particular vantage point, that of the statements of impossibility that occur in mathematics. There is an abundance of such statements: It is impossible for two straight and parallel lines to meet; it is impossible to square the circle; it is impossible that the sum of two even numbers be an odd number; it is impossible to give a proof of the consistency of Zermelo's axioms; and many others.

We might even say that since the phrase "it is impossible that" is simply the negation of the phrase "it is the case that" and that any statement of mathematics that asserts that something is the case can be converted to an impossibility by denying its denial. Thus, "two and two is four" converts to "it is impossible that two and two be other than four." Nonetheless, some statements seem to fit the impossibility format more naturally than others. For example, children are taught to say that "you can't take six from four," and mathematicians say that "a transcendental number is one that cannot possibly satisfy a polynomial equation with integer coefficients." Furthermore, the psychological import of a statement that asserts impossibility is different from one that asserts actuality. ("You mean that such and such is impossible? You mean that no matter what I do, no matter how hard I try, I'll never succeed in...?") There seems to be a time element in such statements. Actuality is here and now, it is complete, but an impossibility seems to bargain with an uncommitted future.

Consider the following three statements:

It is impossible for two integers p and q to exist such that $p/q = \sqrt{2}$.

It is impossible to define the terms in an axiom.

It is impossible to display the first $10^{1,000,000,000}$ decimal digits of $\sqrt{2}$.

The first statement is an old theorem of mathematics. The second is a statement about the foundations of mathematics and the language in which mathematics is written. Since formalized mathematics proceeds by deduction from axioms to theorems, the former are relatively arbitrary starting points and hence indefinable. In the first two statements, the action takes place within the world of the mathematical imagination. In the third statement, if we interpreted the words "to display" in some physical sense, the exterior world and judgments about it now play a role. Isolated, self-contained, wholly formalized mathematics exists only as an idealization; both common discourse and the facts of the exterior world constantly intrude to provide meaning and direction.

My object is to present several statements of mathematical impossibility and to discuss their epistemological status—that is, to discuss what the statements really say and reasons that have been given for believing that they are meaningful and true.

Squaring the Circle

The statement that it is impossible to square the circle is probably the most familiar of all the impossibility statements of mathematics. This statement is very old—the playwright Aristophanes (400 BC) uses the words "circle squarer" as a term of derision. A circle squarer, metaphorically speaking, is a person who persists in trying to do the impossible. Much more narrowly construed, the term "circle squarer" refers to one of a group of people, operating on the fringes of mathematical activity, who believe they have discovered how to square the circle despite what mathematicians tell them. The impossible often exerts a lure that cannot be matched by the possible. The anticipated glee and self-satisfaction of showing that the experts are wrong is considerable.

I begin with a deliberate oversimplification. The problem is this: Given a circle, construct a square whose area equals that of the circle. A second version of the problem is: Construct a square whose perimeter equals the length of the circumference of the circle. The two problems are intimately related; if one is impossible, so is the other, and conversely. What follows here refers to the second formulation.

If you are hearing about the impossibility of circle squaring for the first time, you may say, "The impossibility statement is ridiculous. It goes against common sense. Suppose the circle is a beer barrel, a tire or a trunk of a tree. Just draw a rope around the circumference tightly, snip off the length of rope, measure it, divide the length into four equal parts, and voilà, you have the side of the square whose perimeter equals that of the circle."

If you are not experimentally inclined but are arithmetically inclined and you remember some high-school geometry, you might refute the claim of impossibility this way: "Let us suppose the radius of the circle is 1 foot. I know from my geometry class that the circumference must be 2π feet. I know from my little hand-held calculator that the value of π is 3.1416 (to four decimal places). Therefore, the length of the side of the required square is $2\pi/4$, which computes to 1.5708. Go build a square whose side is 1.5708, feet and you have squared the circle."

These arguments asserting the possibility have brought in three new elements: the circle as a physical object, measurement as a physical act, and construction of the square as a physical act. There is no doubt that we can "square the circle" in either of the ways just mentioned, and our solution would be a good practical solution. But such solutions are open to the criticism that while they provide good approximations, they are not exact in the strict mathematical sense of the word. If our task is to arrive as an ideal mathematical square, residing in the mathematical world, by a sequence of ideal mathematical operations of a certain specified kind, then we must pursue the task totally within the world of mathematical theory. Our experiences with mathematics and physics have led us to two different places—"nature and artifice have collided."

The Greek mathematicians of classical antiquity were rather less interested in going the way of physical measurement than in going the way of pure mathematical theory. It was they who located the problem totally and firmly within the mathematical world, and this placement becomes an essential part of the assertion of impossibility.

We still have a job to do, however, before the problem becomes well posed and leads to an impossibility statement. We must clarify the mean by which we are allowed to "construct" the appropriate line segments.

Geometry is pursued in the real world by both physical and conceptual means. One draws real lines and real curves on a real piece of paper using certain drawing instruments (or, more recently, via computer graphics). The physical environment becomes a laboratory in which constructions, reasoning, and discovery can all take place. The favorite drawing instruments of the Greeks in the classical period were the ruler and compass. With these instruments, the circle and the straight line can be produced. The Greeks also had other instruments for drawing other curves, and discussions of their uses occur in their advanced material. In view of the simplicity of the ruler and compass, the notion of ruler and compass constructions took on a distinguished status. Such constructions lead to figures all of whose parts are built up successively by these two tools. Euclid gave numerous such constructions in *The Elements*.

Now the physical act of construction by ruler and compass is replaced by a formalized mathematical surrogate that states clearly what we are allowed to do. Having done this, the problem of squaring the circle is located in the realm of abstract mathematics. More precisely, with the aid of the algebraization of geometry initiated by Descartes and of the calculus initiated by Newton and Leibniz, our problem is removed from geometry itself and becomes a problem located within the complex of ideas now known as algebra and analysis. To argue about squaring the circle no longer requires that you own a set of drawing instruments, but that you have a profound knowledge of theorems of algebra and of advanced calculus. The problem can now be stated precisely, and the meaning of the impossibility of circle squaring is this: It is a theorem of mathematics that no finite number of algebraic operations of such-and-such a type can lead to the desired result. The truth status of the impossibility is identical to that of any other theorem of pure mathematics, and its acceptance is on that basis.

As a matter of historical fact, in the days when Aristophanes was laughing at the circle squarers, it was not known whether or not such a construction was possible. Repeated failures led the mathematical community to conjecture that it was not. In the course of these failures, many ingenious ruler and compass constructions were devised whose accuracy was very high (but not perfect). Proof of the impossibility was not reached until the 1800s. As stronger assertions emerged, namely, that π is a transcendental number in the sense just defined. For a long time, the proof of this statement was considered to be the outstanding problem of mathematics. Finally, in 1882, the transcendentality of π was established by the German mathematician Ferdinand Lindemann.

Impossibilities within Deductive Mathematical Structures

As we have just seen, to refine and make precise the notion of a mathematical impossibility requires that one confine oneself to a certain limited area of mathematics. Such an area will embrace its own mathematical objects, it will set forth definitions and axioms and ultimately will yield a set of true statements (theorems) relating those objects. This limitation is known as working within a deductive mathematical structure. Historically, the details of the axiomatizations are laid down after a considerable number of results have become clear on a more informal basis.

Here are a few more interesting impossibilities and the areas of mathematics in which the impossibility is now established.

1. Given the edge of a cube, it is impossible to construct by standardized ruler and compass operations the edge of another cube whose volume is twice that of the first. (Galois theory)

2. It is impossible to trisect a general angle with ruler and compass operations. (Galois theory)

Incidentally, duplicating the cube, trisecting the angle and squaring the circle constitute the three classical impossibilities derived from Greek geometry.

3. It is impossible to find a formula that involves only a finite number of arithmetic operations and root extractions, that solves the general quintic equation, or indeed any general equation of degree higher than four. (Galois theory)

4. It is impossible to solve Sam Loyd's "Fifteen Puzzle" in the initial configuration shown here. The Fifteen Puzzle consists of a 4 × 4 board filled with fifteen numbered movable squares or tiles. The problem is to move the squares successively into the empty space until the tiles are arranged in numerical order. (Two-dimensional combinatorial geometry)

1	2	3	4
5	6	7	8
9	10	11	12
13	15	14	

The Fifteen Puzzle

5. It is impossible that there exist more than five types of regular polyhedra: the tetrahedron, the cube, the octahedron, the dodecahedron, and the icosahedron. (Three-dimensional Euclidean geometry)
6. It is impossible to set up a one-to-one correspondence between the set of integers and the set of real numbers. (Cantorian set theory)

The Theater of the Absurd: Conversion of the Impossible to the Possible

It is an error, or, at the very least, it contributes to a misleading view of mathematics, if mathematics is seen as a set of static, formal, deductive structures permanent in arrangement and fixed for all time. A truer picture of the subject is obtained if these structures are viewed as historic but provisional, emerging as new thinking elicits new rules and delineates their scope, and as new creative pressures alter their individual relevancies. The rules may change, hypotheses may change, the order of deduction may be turned upside down to allow hypotheses to become conclusions and vice versa, new interpretations may be found within larger milieus.

Impossible—and yet possible: The $\sqrt{2}$ is the first instance of the absurd in mathematics. Impossible, absurd, within a rigid axiomatic frame of arithmetic to which mathematics was unable to confine itself and still remain creative. The way out was hinted at by much earlier Babylonian mathematics. In a tablet that has been dated to approximately 1700 BC, one finds an excellent hexadecimal approximation of $\sqrt{2}$.

Are there any members x of my structure for which $x^2 = 2$? No, if my structure is the set of rational fractions (i.e., numbers such as 7/5 or 71/86). Yes, if my structure is the set of real numbers. Does the equation $x^2 = -1$ have a solution? No, if x is a real number. Yes, if x is a complex number.

Do you want to divide by zero? Do you want to know what 0/0 is? This "absurd" ratio, when properly interpreted as the limit of legitimate ratios, is the very stuff from which differential calculus can be built. Do you want to know what 1/0 is, i.e., the inverse of 0? Within projective geometry, or within matrix theory, answers to this are given routinely, usefully, and without shame. (E.g., every rectangular matrix has a Moore-Penrose generalized inverse.)

Do you truly want to solve the Fifteen Puzzle given the initial configuration above? Then behave like Alexander of Macedon cutting the Gordian Knot: Lift the little squares out of their frame and into the third dimension. Then arrange the squares properly. This is what you would do if your life depended on getting the desired arrangement.

Mathematical impossibilities are converted to possibilities by changing the structural background, by altering the context, by embedding the context in a wider context. With regard to such conversions, a number of very important questions may be raised.

First, what are the consequences of the act? This question is easily answered by an examination of textbooks that spell out the consequences. Second, why and when does one want to act? While there are numerous instances of the conversion from the impossible to the possible, of which Examples 1 through 6 of the last section are the most famous, to my knowledge there has been no general critique of the act.

Many computer programs ring bells or flash lights when asked to perform something that is disallowed; we are duly warned. When bells ring or light flash, we may indeed want to pay attention. Then again, we may only want to look at the mathematical mechanism that has set the bells ringing. The resolutions of yesterday's shocks often form the bases of today's stabilities.

The Hydra of Impossibilities slain by the Hercules of Context Extension

Further Reading

John D. Barrow. *Impossibility*. Oxford University Press, 1998.

Martin Davis. "The Myth of Hypercomputation." To appear in *Turing Festschrift*, Springer-Verlag.

Philip J. Davis. "When Mathematics Says No." In *No Way: The Nature of the Impossible*, Philip J. Davis and David Park, eds. W. H. Freeman, 1987.

Augustus De Morgan. *A Budget of Paradoxes*. Longmans, Green, and Co., 1872; 2nd ed. 1915.

Mario Livio. *The Equation That Couldn't Be Solved: How Mathematical Genius Discovered the Language of Symmetry*. Simon and Schuster, 2005.

18

Inconsistencies and Their Virtues

The point I'm trying to make is that things which seem inconsistent and even absurd to the lay mind are commonplaces to the mathematical intelligence.
—S. S. van Dine, *The Bishop Murder Case*

Most philosophers make consistency the chief desideratum, but in mathematics it's a secondary issue. Usually we can patch things up to be consistent.
—Reuben Hersh, *What Is Mathematics, Really?*

A Literary Example

To be inconsistent is often thought to be a breach of common sense. But let us see whether in a few words we can come up with some "praise of this folly," as Erasmus said. I am an occasional writer of fiction, engaging in it as an amusement and a relaxation. Compared to professional writers of fiction, I would say that my "fictive imagination" is pretty weak. This doesn't bother me much because I'm usually able to come up with something resembling a decent plot.

I work on a word processor. The processor has a spelling check and a grammar check. The spelling check is useful but occasionally annoying. For example, it does not recognize proper names unless they've been pre-inserted. One time it replaced President Lincoln's Secretary Seward with Secretary

Inconsistent directions?

Seaweed. You know the saying: If something is worth doing, it's worth doing poorly. In these gray days my spelling check has been the source of many laughs. Therefore it would be a pity if it were "improved."

The grammar check in my word processor is only occasionally useful and is mostly annoying. It occasionally throws down a flag when I write a detached sentence such as "London, and Cambridge, so why not Manchester?" When it wants to change a passive formulation into an active one, its suggestion for doing so often ends in a terrible muck. It doesn't catch nonsense. As a test I wrote "The man reboiled the cadences through the monkey wrench," and my spelling and grammar check simply changed "reboiled" to "rebelled."

I recently wrote a long short story entitled *Fred and Dorothy*. What bothered me after I finished was this: Having produced a fairly long manuscript, I was never quite sure whether it was consistent. I don't mean consistent in the sense of pancake batter or whatever the equivalent of that would be in prose style, I mean everyday consistent in terms of time, place, person, etc. Thus, on page 8 of my first draft, I wrote that Dorothy's eyes were blue. On page 65, they're brown. On page 24, Fred graduated from high school in 1967. On page 47, he dropped out of high school in his junior year. On page 73 and thereafter, Dorothy, somehow, became Dorlinda. On page 82, there is an implication that World War II occurred before World War I. On page 34 I wrote "center" while on page 43, I wrote "centre."

I am in good company.

> "You will notice," Pnin said, "that when on a Sunday evening in May, 1876 Anna Karenina throws herself under that freight train, she has existed more than four years since the beginning of the novel. But in the life of the Lyovins, hardly three years have elapsed. It is the best example of relativity in literature that is known to me."
> —Vladimir Nabokov, *Pnin*, Ch. 5, Sec. 5 (paraphrased)

There are inconsistencies in the point of view. *Fred and Dorothy* is a story as told by a narrator; an "I." There is often an "I problem" in fiction, and here's how Peter Gay in *The Naked Heart: The Bourgeois Experience from Victoria to Freud* describes this type of inconsistency:

> Often enough, the narrator [of the first-person novel]—or, rather, his creator—cheats a little, recording not only what he saw and heard, or

was told, but also what went on in the minds of characters who had no opportunity to reveal their workings to him. Most readers, facing these flagrant violations of the narrator's tacit contract with them, suspend their disbelief.

We learn to deal with inconsistencies in books, and we do it in different ways. Suspension of disbelief is only one way. Now, what I would find really useful, if such a thing existed, is a program that checks for consistency as well as my copy editor Louisa does. Louisa is smart. She's careful. She's widely read. She knows my mind. She's worth gold.

Imagine now that I bid Louisa good-bye and replaced her with a consistency checker that I paid good money for. Call the software package CONNIE. Imagine that I run *Fred and Dorothy* through CONNIE. It immediately comes back with a message: "On page 8 you said Dorothy's eyes were blue and on page 65 they're brown. What's the deal?" Did I have to spell out in my text that in the late afternoon October mist, Dorothy's eyes seemed brown to me? How does CONNIE handle metaphor? I wrote: "He saw the depths of the sea in her eyes." Now CONNIE (a very smart package) knew her (its?) Homer and recalled that "Gray-eyed Athena sent them a favorable breeze, a fresh west wind, singing over the wine-dark sea." CONNIE blew the whistle on me: "Hey, wine-dark isn't blue. For heaven's sake, please make up your mind about the color of Dorothy's eyes." Why did Dorothy morph into Dorlinda? That was part of my story: Dorlinda was the name the movie producers decided to give her after she'd passed her screen test with flying colors.

And so on, and so on. Any writer of fiction can explain away *post hoc* what appear to be inconsistencies. In technical lingo, often employed in mathematical physics, explanations that clear up inconsistencies are called interpretations.

I suppose that someone, somewhere, has drawn up taxonomy of textual inconsistencies. It must be extremely long. Mavens who analyze language often split language into three systems with different sorts of meaning: interpersonal, ideational (i.e., ideas about the world in terms of experience and logical meaning), and textual (ways of composing the message). I worry mostly about the first two, and I'd limit my consistency checker to work on them.

Consistent Inconsistencies?

Now let me get to logic and mathematics, to Boole and Frege and Russell and Gödel and Wittgenstein and all those logical fellows. Whereas life does not have a precise definition of consistency, mathematics has a clear-cut definition. A mathematical system is consistent if you can't derive a contradiction within it. A contradiction would be something like $0 = 1$. Consistency is good and inconsistency is bad. Why is it bad? Because if you can prove one contradiction, you can prove anything. In logical symbols,

$$\text{For all } A \text{ and } B, (A \ \& \sim A\ ) \rightarrow B.$$

So, if you allow in one measly inconsistency, it would make the whole program of logical deduction ridiculous.

Aristotle knew about the above implication and had inconsistent views about it. In the literature of logic it's called the ECQ Principle (*Ex Contradictione Quodlibet*), but I call it the Wellington Principle, after the Duke of Wellington, 1769–1852, victor of the Battle of Waterloo. The Duke was walking down the street one day when a man approached him.

The Man: Mr. Smith, I believe?
The Duke: If you believe that, you can believe anything.

Inconsistency is the primal sin of logic. In 1941, in my junior year at Harvard, I took a course in mathematical logic with Willard Van Orman Quine, a man who, in the opinion of some observers, became the most famous American philosopher of his generation. Quine had just published his *Mathematical Logic,* and it was our textbook. The course started just before, or shortly after, the shattering news that J. Barkley Rosser had found an inconsistency in the axiom system Quine had set up. Well, Quine spent the whole semester having the class patch up the booboo in our books, crossing out this axiom and replacing it with that, replacing this formula with that, while we logical greenhorns were anxious for him to get on with it and get to the punch line of logic, whatever that might be. *Q.E.D.* as regards primal sins.

But back to business. If $0 = 1$ and is an inconsistency, then multiplying both sides by 4 gives $0 = 4$. Now is that an inconsistency? It is in Z, but not in Z_4. Come to think of it, how do we know that $0 = 1$ is a contradiction in Z? Perhaps Peano said so, or, if he didn't, I would hope it could be deduced from his axioms about the integers. So, depending on where you're coming from, a set of mathematical symbols may or may not embody an inconsistency, just as in fiction. In fact, a set of truly naked mathematical symbols is not interpretable (or is arbitrarily interpretable). By "naked" I mean that you have no indication, formal or informal, of where the writer is coming from.

Now bring in the famous and notorious Gödel Incompleteness Theorem (GIT). I want to apply it to literary texts. To state it in a popular way, the GIT says that you cannot prove the consistency of a mathematical system by means of itself. If mathematics is part of the universe of natural language, and I think it is, then, with a little thought, I should be able get the GIT to imply that it is impossible to build a universal consistency checker. Or, for that matter and much more important these days, that it is impossible to build a universal computer virus checker.

If a consistency checker can't be produced for mathematics, with its sophisticated and conventualized textual practices and its limited semantic field, then I have serious theoretical doubts about literary texts. CONNIE might catch Dorothy's eyes being simultaneously blue, brown and wine-dark, but there will be some inconsistencies that CONNIE misses.

Mathematics is one way we try to impose order, and we may do it inconsistently. Consider the arithmetic system embodied in the very popular and useful scientific computer package known as MATLAB. MATLAB yields the following statements, from which a contraction may be drawn:

$$1E\text{--}50 = 0 \text{ is true, i.e., } 10^{-50} = 0.$$

$$2 + (1E\text{--}50) = 2 \text{ is false.}$$

Well, we all recognize roundoff and know its problems. And we know, to a considerable extent, but not totally, how to deal with it; how to prevent it from getting us into some sort of trouble. MATLAB and other computational packages embody what has been called non-Diophantine arithmetic.

Is it a contradiction that the diagonal of the unit square exists geometrically but can't exist numerically? At one point in history this was a highly irrational conclusion and one worthy, so the story goes, of slaughtering oxen. Was it a contradiction that there exists a function on $[-\infty, +\infty]$ that is zero everywhere except at $x = 0$, and whose area is 1? It wasn't among the physicists who cooked it up and used the idea productively. It was among mathematicians until Laurent Schwartz came along in the 1940s and showed how to embed functions within generalized functions.

More recently, and in connection with Hilbert's fifth problem, Chandler Davis has written

> I cannot see why we would want a locally Euclidean group without differentiability, and yet I think that if some day we come to want it badly—in which case we will have some notions of the properties it should have—we should go ahead! After five or ten years of working with it, if it turns out to be what we were wishing for, we will know a good deal about it; we may even know in what respect it differs from that which Gleason, Montgomery, and Zippin proved impossible. Then again, we may not...

Thus, inconsistencies can be a pain in the neck, a joy for nit-pickers, and a source of tremendous creativity.

Karl Menger, in his *Reminiscences of the Vienna Circle and the Mathematical Colloquium* tells a story that, in Wittgenstein's opinion, mathematicians have an irrational fear of contradiction. I often agree, but I also realize that mathematicians are smart enough to spirit away a contradiction, as Reuben Hersh has written. Mathematical inconsistencies are often exorcised by the method of context-extension. It is done on a case-by-case basis and is worth doing only after the contradiction has borne good fruit. So the notion of mathematical consistency may be time and coterie dependent, just as in literature.

Logicians who go for the guts of the generic, and who are overeager to formalize everything, have come up with a concept called paraconsistency. There has even been a World Congress to discuss the topic. Ordinary logic, as I have noted, has the Duke-of-Wellington property that if you can prove A and not A simultaneously, then you can prove everything. Paraconsistent logic is a way of not having inconsistency destroy everything. Contradictions can be true. Such a system might be good for certain applications to the real world, where conflicting facts are common.

Walking down the street in paraconsistent London, a man approached the Duke of Wellington.

The Man: Mr. Smith, I believe?
The Duke: My dear Sir, don't let your belief bother you.

When all is said and done, and paraconsistency aside, I don't think I can define consistency with any sort of consistency, but I'm in good company. Paralleling St. Augustine's discussion of the nature of time: Although I can't define a contradiction, I know one when I see one.

In a very important paper written in the mid-1950s, the logician Y. Bar-Hillel demonstrated that unless computers were able to reason about the content of a text, then language translation by computer was impossible. This demonstration dampened translation efforts for a few years. Ultimately it did not deter software factories from producing language translators that have a certain utility and that also produce absurdities. I'm sure that the software factories will soon produce a literary consistency checker called CONNIE, and I will run to buy it; it might be just good enough for me. And if I've paid good money for it, then, as the saying goes, it must be worth it. The absurdities it produces will lift my spirits on gray days and serve to remind me of the "folk theorem" that bad software can often be useful.

Further Reading

Jean-Yves Béziau. "From Paraconsistent Logic to Universal Logic." *Sorites*, No. 12, May 2001.

George S. Boolos, John P. Burgess and Richard C. Jeffrey. *Computability and Logic*, 4th ed. Cambridge University Press, 2002.

Walter A. Carnielli. *Paraconsistency: The Logical Way to Inconsistency*. Marcel Dekker, 2002.

Chandler Davis, e-correspondence.

John W. Dawson, Jr. *Logical Dilemmas: The Life and Work of Kurt Gödel*. A K Peters, 1997.

Peter Gay. *The Naked Heart*. Norton, 1995.

Reuben Hersh. *What Is Mathematics, Really?* Oxford University Press, 1997.

George Lakoff and Rafael E. Núñez. *Where Mathematics Comes From: How the Embodied Mind Brings Mathematics into Being*. Basic Books, 2000.

Karl Menger. *Reminiscences of the Vienna Circle and the Mathematical Colloquium*. Kluwer Academic, 1994.

Chris Mortensen. *Inconsistent Mathematics*. Kluwer Academic, 1995.

S. S. Van Dine [pseud.]. *The Bishop Murder Case*. Scribner's, 1929.

World Congress on Paraconsistency 2000, São Paulo, Brazil. Marcel Dekker Inc., 2002/2004.

19

On Ambiguity in Mathematics

In the literary context, the notion of ambiguity was given prominence in the 20th century by Sir William Empson in his 1930 book *Seven Types of Ambiguity*. Empson defined ambiguity as "any verbal nuance, however slight, which gives room for alternative reactions to the same piece of language" and believed that it was fundamental and an absolute boon to literary creations. Physicist Lisa Randall pointed out that differences in the meaning of words between the colloquial and the professional can cause deep misunderstandings. In particular, she cites the words "relativity" and "theory." In mathematics, many words, such as "positive," "real,"

This is a very old picture. In the blink of an eye, a young lady is converted to an old woman.

and "function." carry specialized mathematical connotations that differ from the colloquial. Mathematics has long had the reputation of being clear, sharp, and totally unambiguous. While this represents an ideal, it is not always the case.

There are, to start with,

1. ambiguities of terms and of notations. Multiple meanings, known as *polysemy*, are frequent. *A* and *B* designate statements. In the logical composition of statements, what does "*A* or *B*" mean? Either

A or *B*, but not both? This exclusive interpretation is not the usual one. Does the symbol *dx* designate the product of *d* and *x* or the differential of *x*? Is log(*x*) the log to the base 10 or to the base *e* = 2.71828...? One and the same symbol may inadvertently be assigned to different quantities or variables within one discussion. The term "independent" may refer to vectors, variables, or random samples.

The adjective "regular" is one of the most overworked words in the mathematical vocabulary. It means different things in different contexts and occasionally within the same context. Example: *Regular polyhedra* sometimes refers exclusively to the five convex Platonic solids and sometimes includes the four non-convex stellated Kepler-Poinsot polyhedra as well. Indeed, the word *polyhedron* itself has a number of different definitions that lead to different objects.

The reverse also occurs; there may be different notations or representations for one concept. This situation is called *polylexis*.

$$1/2 = 1 : 2 = 0.5 = 0.500 = 50\% = 0.49999... = \sqrt{(1/4)}$$
$$= \text{one half} = \text{un demi (in French)}$$

Regular matrix = non-singular matrix = invertible matrix

Eigenvalue = characteristic value = latent value

It is also the case that different notations have their own individual histories, overtones, utility, and implications.

2. ambiguous conclusions, as in the "halting problem." We have been assured that we cannot, by a general procedure, tell in advance whether a randomly produced computer program will halt or will not halt. Either may happen. What is the probability that it will halt? Or consider the so-called butterfly effect in chaos theory: Small variations in the initial conditions of a dynamical system may produce large variations in the long-term behavior of the system and these may not be determinable in advance.

3. alternative options or selections that can be made by a mathematician. Here are a few examples: philosophical or practical acceptance or rejection of non-constructive mathematics; acceptance or rejection of Cantorian set theory; acceptance or rejection of the axiom of choice; adoption of Euclidean or non-Euclidean geom-

etry; acceptance or rejection of proofs via computer. A variety of formalized systems of logics are available.

4. ambiguity of truth values. Mathematical truth cannot be equivalent to provability via axiomatics: There are true but unprovable statements (Gödel).

5. mathematics as metaphor. The process of abstraction allows mathematical concepts to be applied to real-world situations. Thus the concept "four" may be applied to a bunch of bananas or to the number of players in game of bridge. A matrix may be used to describe how the vertices of a graph are connected or to describe a linear transformation of space. A variety of mathematical models is available for each human–physical phenomenon, and such models are often considered as metaphors. You can pick one according to your personal criterion of which is best.

6. changes in concepts and notations over the centuries and millennia: What is a number? What is geometry? What is a function? What is the existence status of mathematical objects? What is an acceptable demonstration? Consider the word "function." Since the writings of Leibnitz in 1694 to the present, the word has meant variously a curve, a formula, a relation, a computer program.

Within "hand-crafted" mathematics, ambiguities create few difficulties because one works within a limited context with its own specialized meanings and notations, and the communication is often within a limited clientele. Problems can arise when one crosses fields. "Surgery" means one thing within medical science and another thing within mathematical topology, and search engines might not distinguish between the two.

The real problem of ambiguity begins with the attempt to automate the processing of natural or scientific languages. When it is a question of machine translation, information analysis and retrieval, text or speech processing, ambiguities are a considerable headache. The current production of nonsensical—often hilarious—material produced by software now available is testimony for the difficulty of the problem. Creating strategies for disambiguation has given rise to a new mathematical–computer science field whose achievements to date have been modest.

Assessing the situation, literary analysts Greig E. Henderson and Christopher Brown have written:

> Ultimately, we must develop a rather complete set of such macros[1] to disambiguate all the notations and usages that we expect to find in the author's materials. These macros would be included in the provided style files, and authors would be encouraged to use them. We do have the advantage, however, of not having to deal with all of mathematics, for which the richness of notations with multiple interpretations becomes quite overwhelming.

To these words I can only add: stay tuned.

Further Reading

Bill Byers. "The Ambiguity of Mathematics." *Proceedings of the International Group for the Psychology of Mathematics Education, Haifa*, Vol. 2, 1999.

Greig Henderson and Christopher Brown. *Mathematical Ambiguities* (http://dlmf.nist.gov/about/publications/nistir6297/node11.html).

Nancy Ide and Jean Veronis. "Word Sense Disambiguation: The State of the Art." *Computational Linguistics*, Vol. 24, 1998.

Lisa Randall. "Dangling Particles." *New York Times*, Op-ed, Sept. 18, 2005.

---

1. A *macro* is a combination of commands, instructions, or keystrokes that may be stored in a computer's memory and can be executed as a single command by a single keystroke or a simultaneous combination of keystrokes.

Mathematical Evidence: Why Do I Believe a Theorem?

I believe in the truth of a theorem for a variety of different and interlocking reasons. I call these reasons mathematical evidence.

Beliefs

1. I believe that 2 + 2 = 4. I have counted two groups of two peaches, and my counting stopped at four. I looked on the back of my little elementary school notebook where an addition table was printed. I have defined 4 to be 2 + 2; I asked a friend; she told me the answer is 4. I have proved it to be the case (after two hundred pages of logical preliminaries) *à la* Whitehead and Russell.

Still, if 2 + 2 = 4, then 8 + 8 = 16 except in "clock arithmetic," where 8 + 8 = 4. The context is therefore important and may be ambiguous. I do not always know when I should add in the manner set out in the elementary addition table. Perhaps more deeply, there is no definitive statement—there can be no definitive statement—as to when it is appropriate to add two integers, in a physical, social or economic context. (See also Chapter 9, "When Should One Add Two Numbers?")

2. I believe that 4519 × 8203 = 37,069,357. I have not done this by brute counting. I worked it out using a method (algorithm) I learned in grade school. I asked my wife to work it out. She did so, and she got the same answer. I checked out my answer using certain tests (e.g., by casting out nines). I realize this is a necessary but not sufficient test. I tried it backward and found that 8203 × 4519 = 37,069,357. My little calculator that I got free when I opened up a bank account gave me the same result. I have not proved the product is correct by going back to axiomatics.

Caution: Let us conduct the following *Gedankenexperiment*. Imagine a little hand-held calculator, used widely in homes for personal finances, etc., which has been gimmicked up so that, with one exception, all the outputs are "normal." When one asks it to compute 8203 × 4519, however, it outputs 37,096,357 instead of the correct value. Now what status can we ascribe to this little calculator? Should it be thrown out immediately?

In the first place, over several lifetimes of steady use, the probability is that this particular product will never be demanded. Secondly, if it were required, then, depending upon the use to which the answer is put, the output, which is strictly erroneous, may be perfectly acceptable. The output, after all, is in error by less than 0.07%, and if we were doing certain engineering computations or a computation relating to with the National Budget of the United States, the discrepancy might hardly be noticed. In terms of utility, then, we would say that the calculator could be given a certificate of high quality.

The mathematics embodied in this particular gimmicked instrument, when considered as an input–output device in its own right, is not erroneous. It is simply producing a different arithmetic obeying different rules and standing in an approximate relationship to the true, or standard arithmetic. It has an integrity of its own. It can lead to contradictions and paradoxes only when interpreted and manipulated in terms of the rules of standard arithmetic, but not when interpreted in terms of its own exceptional behavior, for which special rules might be developed.

3. I believe that the Pythagorean Theorem is true for a right angled triangle. I took a ruler, and I drew a few right-angled triangles and measured the three sides. I found the theorem is approximately true. I proved the theorem in high school. I also learned that there are more than one hundred different elementary proofs known. Some of these proofs make the statement so plain that one can see or feel its truth, geometrically, visually. In calculus and in higher-dimensional geometry, one defines arc length by the Pythagorean theorem. It is the basis of so much mathematics that if it weren't true, lots of books would have to be thrown into the trash.

4. I believe that a negative times a negative yields a positive. By definition, a negative quantity is a certain concept that extends, unifies, and creates new representations, formalisms, and new possibilities of inquiry.

Red ink on an accountant's sheet designates negative quantities. A negative quantity is often justified mathematically by a geometric figure as when ticks on a long straight line indicating the positive integers are extended backwards. Sometimes it is justified by vectors. In more sophisticated developments, a negative quantity is an equivalence class of a pair of positive quantities. The multiplication of negative quantities can be conceptualized as, e.g., the elimination of three debts of four dollars each.

5. I believe that I cannot square the circle. I tried (when I was younger) to square the circle and failed. I have read that many people have tried and failed. I have read that the statement has been proved, but I have not read the proof of the impossibility. I would conjecture that most of my professional colleagues have not read the proof, yet they also believe that it is impossible to square the circle. They waste no time trying. Let me say even more: I believe that the number π is a transcendental number from which it would follow that one cannot square the circle.

And yet there are "cranks" or "unbelievers." People frequently write to universities with constructions that purportedly square the circle. Such people often do not know the technical sense within which the impossibility has been demonstrated. Whom do you believe? Why do you believe the establishment and not a crank? (Georg Cantor was once considered a crank by some very creditable mathematicians.) Is it impossible that the establishment has made an error? But we must go with the agreed upon judgement of the group of informed professionals.

6. I believe that there is an infinite number of twin primes. I have not proved it. I believe that no one has proved it. There is a lot of experimental evidence in its favor. The statement can be given a probabilistic interpretation and a "proof" that fits in nicely with experience.

Question: Does the truth of a mathematical statement depend on someone having proved the statement?

7. I believe that the derivative of $y = x^2$ is $dy/dx = 2x$ and that the process has geometrical and numerical significance. I checked it out using the rules of differential calculus. I checked it (approximately) graphically by drawing tangents to a parabola. I checked it (approximately) computationally using difference quotients. I know that Newton knew this fact. I know that Leibniz arrived at it employing differentials. I also know that George Berkeley (1685–1753) said infinitesimals are nonsense. I know that the calculus was rehabilitated years ago via the theory of limits. I know that infinitesimals were rehabilitated about thirty years ago by the work of Abraham Robinson. I found that the rehabilitation of infinitesimals did not add one iota of confidence to my acknowledgment of the validity of the result.

8. I believe that a certain result in computational hydrodynamics gives an accurate picture of a certain fluid flow. I have had some experience in the field and have confidence in the group that produced the result. I know that what they have computed is of utility in the design of new airplanes. There are also conclusions in flow theory reached through applied mathematics say, in weather, ozone layers, pollution, etc., in which my confidence level is somewhat lower because the mathematical equations (models) on which the conclusions are based may be inadequate.

9. I believe that every square matrix can be reduced to Jordan canonical form. This comes close to my current everyday work in numerical methods. I use this result frequently to prove other theorems. These theorems mesh well with a certain corpus of matrix material that I'm familiar with. I teach this canonical form. I express it in several different ways. I know that there are many different proofs of it in many textbooks. In view of the numerical instability of the Jordan form, I have never followed any of the available proofs, finding them too long, excessively boring, or requiring too much abstract background or specialized language. I have never devised a proof of my own, nor have I ever presented a proof in class. Had I done so, I would have repeated by rote (as is often done) what I have found in textbooks.

10. I believe in the so-called infinity of the positive integers, although I'm not sure what the word "infinity" means. A good deal of what I do mathematically seems to be based on assuming the existence of an infinity of positive integers. (I stress the word "seems.") I have problems with this belief and its consequences. Yet I often act inconsistently with my beliefs.

Is the infinity of the integers a fiction that contradicts common sense? Is it a convenience? An agreement? A synonym for a concept that can be formalized in a finite number of symbols? I know that it is claimed that the whole of mathematics can be formalized and displayed in a finite number of symbols, but where are these symbols?

The set of positive integers is often designated by the symbol N. Do the integers live in the symbol N? Can I display them all? Yes and no. It depends what you mean by the word "display." Ah yes, the infinite set of positive integers is said to live in the world of ideas. Where or what is that? Ask your favorite philosopher, cognitive psychologist, or even psychoanalyst.

Can I deny the statement that there is an infinity of integers? Yes, I can. Some scholars have. What are the consequences? That a certain portion of mathematics becomes meaningless. What would be so terrible about that? For many professional mathematicians and for a variety of reasons, large portions of other mathematicians' work are meaningless to them. Mathematician A might not be able to come to grips with the ontology of Mathematician B.

Components

Having approached the question of mathematical evidence "horizontally," through instances of my own psychology, let me pursue it "vertically" and point out some of its components—not necessarily independent. A complete taxonomy would be impossible, and the list that follows is in no particular order and some items may overlap. I suspect also, that—quite apart from the so-called increased rigor that entered mathematics around 1820—what might have been considered evidence four hundred years ago might not now be so considered. A similar statement might be made for the mathematics as yet unborn. Note also that, as in law, evidence can be mistaken.

Entire books could be written about each of the components of mathematical evidence below, and in some cases have been. For example, the literature on logical proof is enormous. Beliefs based on some evidence and later prove to be false often cause surprise; they have given rise to many books of counterexamples.

1. Axioms. Mathematical discourse gets off the ground (but not in the historical sense) with basic indefinable concepts, definitions, naming, axioms. Thus, a square is a plane figure with four equal sides and four equal angles because we say that's what a square is. Within Euclidean geometry, I accept the parallel axiom which says that through a point not on a given line, one and only one parallel to the given line may be drawn. Two hundred years ago, I wouldn't have added the qualification.

2. Proof. In the current standard mathematical sense, proof is a sequence of assertions and inferences (often abbreviated and parodied as the sequence "definition, theorem, proof"), in which each assertion follows from previous assertions by logical rules of inference. One frequently begins by citing previously established theorems.

Proof serves many functions. It can be a road to understanding, to abstraction, to generalization, and to further discovery. Logical proof is one of the great things that mathematics has and other intellectual disciplines do not, but to me it is just one brick—OK, sometimes the keystone—in the larger house of mathematical evidence.

Examples. The deductive schemes in Euclid; the proofs in "rigorous" mathematics texts and monographs. (See also Chapter 11, "Deductive Mathematics.")

3. Computation. Hand computations or computations made by mechanical or electronic devices may lead us to certain conclusions. This is what scientific computations or computer proofs are all about.

4. Visual evidence.

Example 1: If $T(n)$ is the nth triangular number, i.e., $1+2+...+n$, then $T(n) + T(n + 1)$ is a square number. Put the two "triangles" together, one upside down and one right side up to make a square.

$$T(2) = 3; T(3) = 6; T(2) + T(3) = 9.$$

o		oo		ooo
oo	+	o	=	ooo
ooo				ooo
6	+	3	=	9

Example 2: If one angle of a triangle and its area remain constant, then if one side adjacent to the angle increases the other adjacent side must decrease. (Draw a figure and you'll soon see why.)

Example 3: The "Swiss cheese": One cannot exhaust the area of a circle by plastering it with a finite number of non-overlapping smaller circles. There will always be some area left over.

5. Other physical evidence.

Examples: Find the volume of a bottle by pouring in water and measuring it against a standard volume. Find the eigenvalues of a certain differential operator by observing the tones of a kettle drum. Legend has it that Gauss triangulated the peaks of three mountains near Göttingen in an attempt to find whether the angles of a triangle sum to 180 degrees.

Comment: Archimedes used physical reasoning and also wrote against it. Mixed modes or methods (i.e., methods that mixed mathematics and physics) were called *metabasis*. There was a fear or dislike of metabasis because it violated canons of intellectual purity. (See Chapter 22, "The Decline and Resurgence of the Visual in Mathematics," for more discussion.)

6. Analogy.

Example: The infinity of prime numbers and the examination of a long list of primes suggests that there is an infinity of twin primes (e.g., 11 and 13 or 17 and 19). But there is much more substantial evidence in favor of this conjecture. It is even conjectured, and with cogent reasons, that the number of twin primes less than the number x, $\pi_2(x)$, is approximately equal to

$$\pi_2(x) \sim 2\Pi_2 \int_2^x \frac{dx}{(\ln x)^2},$$

where Π_2 is a constant, dubbed the twin prime constant, and whose value is 0.660161... A rigorous proof of these assertions is yet to be found, however.

Something that is true in two dimensions suggests (often mistakenly) that a similar statement may be true in three dimensions. The three altitudes of a plane triangle always meet in a single point. The four altitudes of a solid tetrahedron do not always meet in a single point. This has been known since at least 1827, by Jakob Steiner.

7. Formalism, symbol manipulation. The existence of an interesting but unusual formal expression derived by formal manipulation is often evidence that a cogent theory is lurking behind. This principle often goes under the rubric "the symbols are often wiser than we are."

Example: The equation $1 - 1 + 1 - 1 + ... = 1/2$, which, *prima facie*, seems to be nonsense, is validated within the extensive theory of divergent series.

8. Aesthetics, symmetry, pattern, a sense of rightness, a sense of things "fitting together properly."

Example: Two angle bisectors of an isosceles triangle have equal length. The reverse is also true: if two angle bisectors of a triangle have equal length, then the triangle must be isosceles.

9. Agreement for utility.

Example 1: Minus times minus equals plus. Elementary algebra texts provide strong instances of the utility of this agreement.

Example 2: The standard truth tables for the logical connectives. Thus, the truth table for the conjunction of two statements A and B reads as follows:

A	B	A & B
T	T	T
F	T	F
T	F	F
F	F	F

10. The Pólya heuristic. One form of this is the following: If A implies B, and if B is true, then this makes it more likely that A is true. Example: lots of people go to the beach when the weather is good. If I know only that lots of people have gone to the beach, it makes it likely that the weather was good.

11. Experimentation, trial and error, induction. That something works for $n = 1, 2, 3$ is evidence that it may work for all numbers. If, by experimentation, we have found that it works for $n = 1, 2, 3, ... 1,000,000$, the evidence becomes stronger. If not, and if it works only for some cases, it may be important to identify these cases.

12. "Quasi-proofs," statistical or probabilistic evidence, random testing. If something works in a specific but random case, it is evidence for its general truth. If a polynomial $p(x)$ is zero for one value of x, is it identically zero? Not necessarily. But if a polynomial with rational coefficients is zero for one transcendental value of x, it is identically zero.

13. Common sense, naïve mathematical experience.

Examples: Old sayings, like "Figures can't lie." The shortest distance between two points is a straight line. The rigid motion of an object preserves lengths, angles, areas, volumes of the object. You can always add 1 to a number and get a greater number.

14. Algorithmic induction, belief in the stability of input–output processes. If input A has yielded output B once, then that is evidence that input A will always yield output B. This is the basis of our confidence in computers.

15. Utility. If a process works in some sense (frequently physical), then it must be correct. There is no need to fix it and little need to prove it.

16. Imaginative hypotheses.

Examples: Imagine that there are integers p and q such that $p/q = \sqrt{2}$, then use a contradiction derived from number theory to show the contrary. Cantor's diagonal process, in which we are asked to imagine that all the real numbers have been put in a list, one after another, and that we can argue logically about this list.

17. Insufficient reason: Buridan's Ass. The 14th-century French philosopher Jean Buridan argued that a perfectly rational ass placed exactly between two equal piles of hay wouldn't have sufficient reason to go to one or to the other. Hence the poor animal would starve.

Example: If you pick a point in the Cartesian plane at random, the probability it lies in the upper half plane is $1/2$. Of course, this hinges on what you mean by random.

18. Lack of contradiction so far. Well, we haven't gotten into trouble so far. And if we have, we can always patch it up, somehow.

Example: Read Imre Lakatos' book on how the Euler-Poincaré theorem was patched up over a period of centuries. Is it now, at long last, stated and proved correctly? What gives mathematicians the confidence that it has been?

Comment: I once asked a very distinguished historian of mathematics whether mathematics was true. His answer: "Of course it isn't. Mathematics is simply an instance of a self-justifying system." I came back with "Like psychoanalysis, where if you don't believe it, you must be crazy?" The historian answered, "I wouldn't know about psychoanalysis."

19. Intuition, mysterious hunches. Note that intuition and hunches may go both ways simultaneously: The feeling that such and such is true together with other feelings that such and such is false.

Example: George Bernard Shaw remarks in the introduction to his play "Androcles and the Lion," "People believe not necessarily because it's true but because in some mysterious way it catches their imagination."

20. Compulsion, indoctrination, memory. Alfred North Whitehead pointed out somewhere that one of the goals of mathematics is to eliminate the necessity of constant rethinking.

Example: I learned the times table. It says $9 \times 8 = 72$. I use that answer without further ado. My mind doesn't search for the reasons that underlie this answer.

21. Authority.

Examples: The textbook said so. The table of integrals said so. The computer said so. The genius said so. The mathematical culture or establishment said so.

Apropos of authority, here is a nice quote from Jeremy Gray's article. (See also the Further Reading in Chapter 22, "The Decline and Resurgence of the Visual in Mathematics.")

Jacobi [famous German mathematician (1804–1851)], in a letter to Alexander von Humboldt, wrote that "If Gauss says he has proved something, it seems very probable to me; if Cauchy says so, it is about as likely as not; if Dirichlet says so, it is certain."

Comment: If researchers had no faith in what they learned from various sources, they would be most severely hampered in their work. If, on the other hand, researchers had absolute faith, there would be no revolutionary progress.

22. Theology.

Examples: God's word, God's construction.

This feeling is very old and never totally discarded. Leopold Kronecker stoked it up a century ago: "God created the integers and all the rest is the work of man." This is often turned around: Mathematics represents order, and order is evidence of God. (See, e.g., Hermann Weyl, as cited in the Further Reading section below.) Srinivasa Ramanujan claimed that his Goddess told him certain mathematical things.

In the examples above, I have given many different reasons why I believe that a statement of mathematics is true: proof, induction, intuition, definition; sense, experience, and experimentation in the physical world; experimentation and trial within the world of pure mathematical operations; applicability to the social worlds; utility, computation, sense of appropriateness when mathematical ideas fit together; aesthetic sense, convenience, authority; and there are numerous others. If I were living centuries ago, I probably would have added applicability to the transcendental, the hermetic or religious world.

Belief is a part of the process of mathematical enculturation, a part of the social contract that enables us to interact and communicate with one another expeditiously and profitably along certain lines. In the final analysis, what makes a mathematical statement convincing is a selection from all of the above.

Further Reading

Dana Angluin. *How to Prove It* (http://www.cl.cam.ac.uk/~jeh1004/mirror/maths/how-to-prove-it.html), circa 1984.

Jonathan Borwein and David Bailey. *Mathematics by Experiment*. A K Peters, 2004.

Philip J. Davis. *Mathematical Evidence*. First Alfred North Whitehead Lecture, Imperial College, London, 1998.

Reuben Hersh, ed. *Eighteen Unconventional Essays on the Nature of Mathematics.* Springer-Verlag, 2005.

Gila Hanna. "The Ongoing Value of Proof." *Proceedings of the International Group for the Psychology of Mathematics Education,* Vol. I, 1996.

Hans Havlicek and Gunter Weiss. "Altitudes of a Tetrahedron and Traceless Quadratic Forms." *American Mathematical Monthly,* Vol. 110, 2003.

Imre Lakatos. *Proofs and Refutations.* Cambridge University Press, 1976.

Roger B. Nelsen, ed. *Proofs without Words: Exercises in Visual Thinking, Vol. II.* Mathematical Association of America, 1993, 2000.

Charles S. Peirce. "The Fixation of Belief." *Popular Science Monthly,* Nov. 1877 (available on the Internet).

George Pólya. *How to Solve It: A New Aspect of Mathematical Method.* Princeton University Press, 1945.

George Pólya. *Mathematical Discovery: On Understanding, Learning, and Teaching Problem Solving.* Wiley, 1962.

Sir Peter Swinnerton-Dyer. "The Justification of Mathematical Statements." *Philosophical Transactions of the Royal Society,* A. Vol. 363, No. 1835, Oct. 2005.

Hermann Weyl. *The Open World: Three Lectures on the Metaphysical Implications of Science.* Yale University Press, 1932.

21
Simplicity, Complexity, Beauty

I cannot tell by what logick we call a toad ugly.
—Sir Thomas Browne, *Religio Medici*, 1643

What is simplicity?

People answering e-mail often interlace their replies with the lines of the letter received. This is an easy thing to do. If such a letter then goes on to a third party, there may occur the interlacing of two different people. I often find such communications confusing, and it takes me time to unlace. What is simple for me may not be simple for you. Is trigonometry simple? To college professors of mathematics it is. Thus simplicity may depend on who you are and where you are coming from. It is simpler (in a certain sense) to take a shortcut across the park on an unmarked path. But is it socially desirable to do so? Can simplicities be compared? Is it easier to roll out of bed than to boil a four-minute egg?

Philosopher William of Ockham told us to achieve simplicity by getting rid of unnecessary concepts (Ockham's Razor). But what is unnecessary in a complex world? Theories of Everything (TOEs) try to simplify the cosmos despite constant warnings of oversimplification. There has been much discussion of simplicity in art and design and simplicity in prose.

Architect Frank Lloyd Wright wrote "A thing to be simple needs only to be true to itself in an organic sense." (March 1901) The Romanian sculptor Constantin Brancusi said something similar: "We are not seeking simplicity in art, but we usually arrive at it as we approach the true nature of things."

"To thine own self be true," said Polonius, but short of a hundred hours on a shrink's couch, and perhaps not even then, do we know what our own true nature is?

Do the famous squares of the Russian artist Kasimir Malevitch (1878–1935), as simple a piece of graphics as one might imagine, get close to the true nature of things? Artists themselves rarely give a quantitative definition of what they mean by simplicity, but the mathematical mind, in contrast, seeks definitions in terms of its own raw materials, for example, quantification, symmetries, compact formulas, brief proofs. Seeking the divine in the proportions of the human body, some Renaissance artists, including Michelangelo, sought to quantize these proportions. In 1933, the mathematician George D. Birkhoff proposed a mathematical aesthetic measure as a relationship between order and complexity. These measures, like measures of the quality or desirability that are found in, among others, the annual list of what colleges are best, have not been plucked at random from the heavens. Although there is usually some reasoning and experience behind their formulation, they must be taken with more than one grain of salt.

Many mathematical measures of the simplicity or the readability of works of prose have been proposed. For example, in the 1940s, Rudolf Flesch proposed a measure based on the average number of words per sentence and the average number of syllables per word:

$$M = 206.835 - (1.015 \times \text{ASL}) - (84.6 \times \text{ASW}),$$

where M is the readability, ASL is the average sentence length, and ASW is the average number of syllables per word.

More complicated words and longer sentences yield lower values of M and hence imply low readability. A high value of M implies ease in understanding. This seems like common sense: Comic books come in with an index of 95, the New York Times, 39; Lincoln's Gettysburg Address comes in at 62.8. The IRS tax code, I would suppose, gets you into the minus quantities. One caveat: Flesch's formula is tailored for English; don't use it on philosophical German.

Some of the proposed criteria for readability have been built into computer scanners for spelling and grammar. Microsoft Word has such a scanner. Thus, the previous words of this essay have a Flesch readability index of 55.7, which translates into a Flesch-Kinkaid school grade level of 11.2.

Now let's talk about beauty in mathematics. In mathematics, beauty is often linked to simplicity and simplicity to truth. The aesthetic component in mathematics is a strong inner driving force for research. If the beautiful is perceived as useless, publicists for the public support of mathematical research argue for the "usefulness of the useless" by pointing out that all mathematics is potentially useful, even though the gap in time between the development of new mathematics and its application may be great.

The aesthetics of mathematics has been discussed by many authors. I find an article by Gian-Carlo Rota particularly thoughtful. Ignoring empirical corroboration and evading the question of importance, Rota's article concentrates on the mathematical sequence: definition, theorem, proof. He asserts that beauty can be found independently in each of these. There are beautiful definitions, which, once made, can immediately inspire a whole industry of investigations. There are beautiful theorems with ugly proofs, one of which is the prime number theorem, which provides an increasingly accurate figure for $p(n)$, the number of primes less than a given number n. Picard's theorem asserting that an entire analytic function takes on all values with two possible exceptions has a most beautiful proof: five lines.

What are some of the hallmarks of mathematical beauty? Thinking of the proof process, Rota mentions simplicity, a brilliant step, fruitfulness, revelation. "A proof is beautiful," Rota wrote, "when it gives away the secret of the theorem, when it leads us to perceive the actual and not the logical inevitability of the statement that is proved."

What are some further features cited by Rota? Beauty does not admit degrees of comparison. (Tell that to the promoters of the Atlantic City beauty contests!) To state, for example, that the prime number theorem is more beautiful than Picard's theorem would be nonsensical. Beauty should not be confused with elegance, the latter often residing in mere presentation. The perception of beauty does not occur in a flash. To think that it does is what Rota calls the "light bulb error": "The appreciation of mathematical beauty requires thorough familiarity with mathematics, and such familiarity is arrived at the cost of time, effort, exercise, and *sitzfleisch*."

As a consequence, the beauty that the educated public finds in mathematics—if, indeed, it finds beauty at all—is different from the beauty that professional mathematicians find. I would emphasize—and this would be close to the position of the aesthetes of the late 19th century—that the perception of the aesthetic component in mathematics may be confined to an elite.

Rota also makes several claims that I find contradictory. He says that beauty is objective even as truth is objective. At the same time, he asserts that the perception of beauty depends upon particular schools and periods of history. Returning to the necessary long experience that limits the general perception of mathematical beauty, Rota believes that teachers' attempts to arouse interest in mathematics on the basis of beauty are bound to fail. This position may be compared with the statement of artist James McNeill Whistler in his *Ten O'clock Lecture* that art is "selfishly occupied with her own perfection only—having no desire to teach."

Has simplicity and hence beauty in mathematics been itself mathematized or quantized? Logician Jan Mycielski wrote me about certain tentative attempts to relate simplicity to complexity and also to the sense of being convinced. He distinguishes logical complexity from mathematical complexity and refers both to the lengths of proofs in their structurally atomic formats. How can a given sentence be replaced by a logically equivalent shorter sentence? The formalization of such a process would be "a first step in the program of building a machine into which we could feed our mathematical knowledge and which would be better than us in proving our conjectures."

Mathematical simplicity should not be confused with mathematical utility. What is useful mathematically is not necessarily beautiful, and the assertion that the beautiful should be useful has been decried by such aesthetes as Whistler, Oscar Wilde, and the famous mathematician G. H. Hardy. The useful is often ugly. In numerical methods, the best codes (e.g., for the various matrix operations) are often long, complex, and company confidential. In short, they are polyalgorithms combined with rules of thumb and are not normally explained in detail in courses in numerical methods.

Let me conclude with a few observations on simplicity in physics, an area in which I've had little hands-on experience.

Einstein believed that the true laws of nature are those that have the simplest mathematical or aesthetic formulations. Physicist David Park believes that simple theories (short formulas, few arbitrary elements) are most likely to be true. Yet, he points out that for 25 years particle physics has been dominated by the "standard model." It agrees with every experiment, yet it has something like 17 arbitrary constants, and the whole thing has such a patched-together look that nobody believes this is its final form. An interesting possibility is that a final simple form may require a kind of mathematics that is not yet invented, something completely new, like the noncommutative algebra of 150 years ago. And if it hadn't been for the Ricci-calculus, Einstein would not have been able to arrive at the General Theory of Relativity.

But there are contrary opinions.

I once heard [Paul] Dirac [British physicist, 1902–1984] say in a lecture, which largely consisted of students, that students of physics shouldn't worry too much about what the equations of physics mean, but only about the beauty of the equations. The faculty members present groaned at the prospect of all our students setting out to imitate Dirac.

—Steven Weinberg, *Towards the Final Laws of Physics*

Whether beauty in theories is a reliable indicator of proximity to the truth depends ultimately on empirical facts about the world: I shall argue that the evidence so far is negative. I shall conclude with some remarks about the aesthetic preferences that have played a conservative role in scientific revolutions.

—James McAllister, *Beauty and Revolution in Science*

Nonetheless, philosopher McAllister advances two major theses. The first is that aesthetics, or, more precisely, aesthetic induction, plays an important and rational role in the creation and acceptance of physical theories—and this despite what is commonly admitted to be an uncertain correlation with experience. The second thesis is that revolutions in science are essentially "aesthetic ruptures" and that these ruptures can be given a rational basis.

Rational? Aesthetic? These words do not appear in Percy W. Bridgman's (recipient of the Nobel Prize in physics, 1946) description of scientific breakthroughs:

What appears to him [i.e., the scientist] to be the essence of the situation is that he is not consciously following any prescribed course of action, but feels complete freedom to utilize any method or device, whatever which in the particular situation before him, seems likely to yield the correct answer. In his attack on his specific problem, he suffers no inhibitions of precedent or authority, but is completely free to adopt any course that his ingenuity is capable of suggesting to him. No one standing on the outside can predict what the individual scientist will do or what method he will follow. In short, science is what scientists do, and there are as many scientific methods as there are scientists.

—*Reflections of a Physicist*

When the fundamental structures of the universe appear to have been modeled in beautiful mathematics, we are faced with a profound mystery that some will merely accept while others will want to explain.

Further Reading

Percy W. Bridgeman. *Reflections of a Physicist*. Philosophical Library, 1950.

Michele Emmer. "Aesthetics and Mathematics: Connections throughout History." In *Aesthetics and Computing,* Paul Fishwick, ed. MIT Press, 2005.

Lynn Gamwell. *Exploring the Invisible: Art, Science and the Spiritual*. Princeton University Press, 2002.

James McAllister. *Beauty and Revolution in Science*. Cornell University Press, 1996.

Jan Mycielski, e-correspondence.

David Park, e-correspondence.

Gian-Carlo Rota. "The Phenomenology of Mathematical Beauty." *Synthèse*, Vol. 3, No. 2, May 1997.

The Decline and Resurgence of the Visual in Mathematics

The following anecdote has been told about numerous mathematicians. The algebraic geometer Oscar Zariski (1899–1986) liked pictures. But pictures can be misleading, and Zariski worried lest they only reveal special cases, and the whole school of Italian geometry in which he had been trained might as a consequence go haywire. In presenting proofs in class he used to draw little pictures in the corner of the blackboard to help him recapture the heart of the matter, and then erase them rapidly as though they were polluting and he were ashamed of having drawn them.

There is visual or geometric sensibility and understanding and there is analytic or formulaic understanding. I will discuss the historic relation between the two in mathematics and its applications. A discussion of the cognitive psychology of these two types of understanding is beyond this chapter.

To give an initial and very simple description of the two types, here is an example.

The analytic statement of the Pythagorean Theorem is

$$C^2 = A^2 + B^2, \qquad (*)$$

where C is the length of the hypotenuse of a right triangle and A and B are the lengths of the legs.

The Pythagorean Theorem, as presented by Euclid, is elucidated by the diagram on the next page.

Euclid's diagram is geometric. It shows the square outside the triangle and not overlapping. The accompanying proof would have to be modified if the squares overlapped. Euclid never multiplied geometric

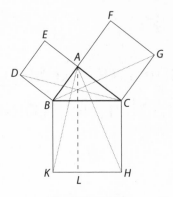

Euclid's proof

magnitudes together: He worked with lines, not with lengths. Although he worked with equality and with "doubles," as Ivor Grattan-Guinness pointed out, he presented a quantitative geometry without arithmetic.

Geometry

In the period of classical Greek mathematics and in the late Renaissance, there was much interest—almost an obsession—on the part of mathematicians, artists, metalworkers, astronomers, and mystics to produce polyhedral constructions that exhibited many symmetries. Engravings and models were created, theorems proved, and applications made in a splendid and often amazing mix of art and mathematics that hardly recognized boundaries between the two.

The distinction between art and mathematics was of little importance. The unity of the arts and sciences was symbolized in the title page of Jamnitzer's *Perspectiva* that bore symbolic figures of four of the liberal arts—arithmetic, geometry, perspective, and architecture. Both Leonardo da Vinci and Michelangelo sought the ideal mathematical proportions of the human body as an expression of the divine blueprint.

Projective geometry has its beginnings in the development of Renaissance painters of perspective. It was followed later by the observation (and proofs) of simple but striking point–line phenomena such as Rene Desargues' and Pappus' (of Alexandria) theorems.

However, by 1687, when Sir Isaac Newton published his *Principia*, although most of the propositions and theorems were illustrated by fig-

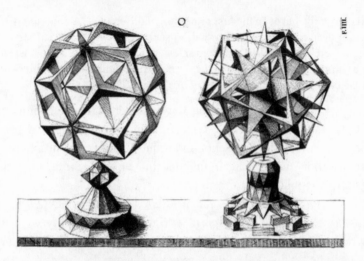

A confluence of art and mathematics (from Wenzel Jamnitzer's
Perspectiva Corporum Regularium, 1568)

ures, very few illustrations displayed what was going on in any graphically lucid way. By that time, the impact of Descartes' *Discours de la méthode* (1635) had moved mathematical discourse toward reducing geometry to algebra. As David Park put it, "One feels that inside every geometric proof there is an analytic proof screaming to be let out." A few illustrations, of course, illustrate vividly what is going on, but rigor and generality were to be drawn from algebra and not from geometry, for geometry was perceived as contaminating the purity of reason.

During the following century, the move toward symbolic abstraction and away from visual representation continued and accelerated. Leonhard Euler, writing in 1755 in the preface to his *Institutiones calculi differentialis*, made two remarkable observations about the nature of differential calculus. First of all, he explicitly rejected geometrical confirmation as a means of testing the validity of calculus, namely, he refused to accept proofs of calculus's correctness based solely on the fact that calculus reached the same conclusions as elementary geometry: Calculus cannot have its own foundation in a geometrical reference. He then observed

I mention nothing of the use of this calculus in the geometry of curved lines: that will be least felt, since this part has been investigated so comprehensively that even the first principles of differential calculus are, so to speak, derived from geometry and, as soon as they had been sufficiently developed, were applied with extreme care to this science. Here, instead, everything is contained within the limits of pure analysis so that *no figure is necessary to explain the rules of this calculus.* [emphasis added]

Joseph Louis Lagrange, writing 30 years later, made the famous statement "*On ne trouvera point des figures dans cet ouvrage,*" and goes on to say "The methods that I expound require neither constructions, nor geometric nor mechanical arguments, but only algebraic operations, subject to a regular and uniform course."

As Ivor Grattan-Guinness put it: "Its 512 pages contained no diagrams, as he stressed in his preface, rigour and generality were to be drawn from algebra uncontaminated by geometry. The book would have been better entitled *Méchanique Algébrique.*"

In the early 19th century, the development of non-Euclidean geometry—in the work of János Bolyai, Nikolai Lobachevsky, Bernhard Riemann and others—undermined the view, derived largely from limited sense experience, that Euclidean geometry has *a priori* truth for the universe, that it is the model for physical space. In the late 19th century, Felix Klein's (1862–1943) *Erlanger Programm* made algebraic group theory a preferred way of viewing geometry. A bit later, David Hilbert and others exposed the incompleteness of Euclidean geometry as a logical structure and the limitations of the visual field with the two or three physical dimensions that form the natural backdrop for visual geometry as opposed to the great generality of geometries that are possible abstractly (finite geometries, complex geometries, symplectic geometries, non-commutative spaces, etc.) when geometry has been algebraicized. Hilbert's axiomatization of Euclidean geometry removed the last vestiges of the visual from the subject.

What also became clear around the early 1900s was the limitation of the eye in its perception of mathematical "truths" (e.g., the existence of continuous everywhere non-differentiable functions, optical illusions, "impossible objects," suggestive but misleading special cases, erroneous constructions):

Whenever geometry has to be really deductive, the process of inferring must be independent of the meaning of the geometrical notions, as well as of the figures. The only things that matter are the relations between the geometrical notions such as have been established in the usual theorems and definitions.

—Moritz Pasch (1843–1930)

Because intuition turned out to be deceptive in so many instances, and because propositions that had been accounted true by intuition were repeatedly proved false by logic, mathematicians became more and more skeptical of intuition... Thus, a demand arose for the expulsion of intuitive reasoning and for the complete formalization of mathematics.

—Hans Hahn (1879–1934, mathematician and member of the famed philosophical Vienna Circle, *Wienerkreis*)

We see from the image below that the special nature of a diagram can lead to false conclusions.

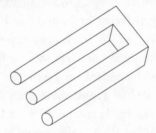

An impossible three-dimensional object

I have the following quip from Peter Lax, who got it from Stanislaw Ulam. "Henri Poincaré said that geometry is the science of correct reasoning from incorrectly drawn figures. Ulam added that his old instructor in *Darstellende Geometrie* reversed it and said that geometry was incorrect reasoning from correctly drawn figures."

The metaphysical claim that the essence of rationalism is to be found in the verbal and symbolic. Today, in the words of David Corfield, we have the following definition of Euclidean geometry:

Euclid's *Elements* is the study of a case of n-dimensional geometry, of the properties of the principal bundle $H \to G \to G/H$, where G is the Lie group of rigid motions of Euclidean n-space, H is the subgroup of G fixing a point designated as the origin, and G/H is the left coset space.

This is very remote from an intuitive, visual understanding of what geometry is all about.

As geometry moved from the visual to the abstract, the movement was epitomized by the war cry of the Bourbakist Jean Dieudonné: "Down with Euclid! Death to the triangles!" (c. 1960). The use of an image was restricted to the heuristic or the pedagogical. The visual image went into a tailspin from which it is only now recovering; its decline was surely a degradation of common sense or at least a denial of its relevance

I have discussed this in the case of geometry, but algebra and the calculus were similarly affected.

Algebra

Early algebra was cast in geometric language, and there was controversy as to whether algebraic manipulation constituted proof. This is reminiscent of today's controversy as to whether manipulation by computers constitutes proof. (Mixing brain and hand work with computer work would be an instance of Aristotelian metabasis.)

A splendid paper by Brandon Larvor points out that Cardano's (1545) conception of rigor was dependent on geometric intuition. But in the space of half a century, we had a movement from the image to the symbol. Thomas Hariot (1560–1621, and the first important mathematician in North America) said that the manipulation of symbols constituted proof. This shift in the notion of proof undoubtedly raised much controversy. Was this change compatible with rigor? And this question became a part of an intense contemporary debate about the philosophical foundations of scientific knowledge. One may ask whether the controversy arose from the feeling that the shift contradicted common sense.

Calculus

The early development of calculus (17th–18th centuries) made great use of pictures. But shortly thereafter there was a backlash in which calculus attempted to assert its independence of geometry. In the early 19th century, the greatest accolade that could have been accorded one mathematician by another was to have called him a geometer. The irony is that at the very time that this honorific was in use, the reasons that called it into being were almost dead. The title was a splendiferous archaism.

One of the best descriptions of the situation with calculus is given in a long paper by Giovanni Ferraro. Here are a few clips from that paper.

During the eighteenth century, calculus abandoned the use of figures as a part of the proof and became a theory which, in principle, could be reduced to a solely linguistic deduction. Eighteenth-century calculus was substantially a non-figural geometry. Calculus was the science of abstract quantities, while figured quantities (that is, quantities represented by a geometric figure) were dealt with by geometry. Calculus achieved its aim of non-figural geometry by basing itself on the notions of variable and function although these notions were completely different from modern ones.

The new conception of calculus as an "intellectual" system was only the beginning of a lengthy evolution that had profound consequences on the whole of mathematics, the very nature of synthetic geometry was to change and become non-figural; that is, it too became a merely linguistic deduction in which the figures were reduced to constituting simply pedagogical support.

Freeing mathematics from every figural consideration and reducing it to linguistic deduction is a complex operation because the absence of the figure makes it impossible to use reasoning that could be entrusted to visual inspection. The reference to the figure must be substituted by adequate axioms and punctual deductions, otherwise lacunae in the deductive process become evident. One merely has to think of the intersection of circles; as long as one can reason on the basis of an inspection of a figure, there is no need for axioms that guarantee the existence of points of intersection; but, whenever the figure is no longer considered an integral part of the deduction process, that lack of an axiom that guarantees the existence of an intersection emerges as a lacuna in mathematical argumentation. Leibniz criticized an early construction of Euclid for just this reason.

Eighteenth-century mathematicians made considerable use of semantic interpretations of the objects of calculus and did not yet feel the real need to resolve certain lacunae in the deductive process. It appeared completely obvious to them that a function was continuous or differentiable; they do not seem to have contemplated the idea that calculus could deal with discontinuous and non-differentiable functions in the modern sense. The lacunae in Euler's and in Lagrange's presentations were noticed by mathematicians at the start of the nineteenth century, such as Bolzano, who [in 1817] wrote:

The most common kind of proof depends on a truth borrowed from geometry, namely, that every continuous line of simple curvature of which the ordinates are first positive and then negative (or conversely) must necessarily intersect the x-axis somewhere at a point that lies in between those ordinates.

There is certainly no question concerning the correctness, nor indeed of the obviousness, of this geometrical proposition. But it is clear that it is an intolerable offense against correct method to derive truths of pure (or general) mathematics (i.e., arithmetic, algebra, analysis) from considerations which belong to a merely applied (or special) part, namely, geometry...

Nonetheless, despite this criticism, the image persisted in calculus for didactic uses.

Metabasis is the classic word for the mixing of modes or categories. Examples of such mixing: figures and analytics, mathematics and mechanics, mind and matter, stasis and change, etc. Avoidance of metabasis is related historically to ancient strictures about purity. It goes beyond scientific inquiry; consider, for example, the separation of powers under the U.S. constitution.

[According to] Aristotle, mathematics described the structure of the world, not its processes. A mathematical science of change was a downright category mistake.

—Amos Funkenstein (See also Steven Livesey.)

Bolzano's moral was clearly to avoid metabasis. In the following quotation Bertrand Russell expresses the desire to avoid metabasis within an already de-visualized form of geometry.

The true founder of non-quantitative geometry is von Staudt (1847). But there remained one further step before projective geometry could be considered complete, and this step was take by [Mario] Pieri [around 1900]... and dealt projectively with a continuous space... Thus, at last the long process by which projective geometry has purified itself from every metrical taint is completed. [1903, 1938, *The Principles of Mathematics*]

Note the words "purified" and "taint." The striving after methodological purity as an aesthetic and creative principle within mathematics has been noted by a number of authors (see John W. Dawson) and is a topic that could stand rather more discussion than it has received.

Visualization

The Greek root of the word "theorem" is *theorein* ($\theta\varepsilon\omega\rho\varepsilon\iota\nu$) meaning "to observe."
—From the *American Heritage Dictionary*

During most of the 19th and 20th centuries, mathematicians neglected, discarded, and even debased the use of images in mathematics. By the late 1800s, when mathematicians produced continuous, non-differentiable, space-filling curves, etc., objects that defied absolute visualization, they were almost ready to pronounce that the human eye was an inadequate reporter and a pitiful liar, because it couldn't accommodate what the symbols were able to produce. This is not to imply that there were no mathematicians who worked fruitfully with the image. H. S. M. Coxeter (1907–2003) and Branko Grünbaum, whose studies embraced the theory of polyhedra, polytopes crystallography, and abstract algebra, filled their pages with numerous figures.

During the years I was a student at Harvard, the mathematical plaster models in the display case in the Department of Mathematics gathered dust. For some time thereafter, I maintained an anti-visual attitude, but, after a few years, I came to the defense of visualization. My interest was spurred by my theoretical work in interpolation and approximation theory and, more generally, in numerical analysis. I came to feel that neither words nor mathematical symbols could capture entirely the total *gestalt* perceived by the eye and brain. In the words of semioticist Michael O'Toole,

We often find that transforming our visual perceptions to words both loses touch with the perceptual reality and freezes our complex responses into a sort of false coherence dictated by the structures [and limitations] of our language.

In 1974, three years before Benoit Mandelbrot published his pathbreaking fractal book, years before computer graphics came into its present glory, I suggested that our view of mathematics be widened to include what I called "visual theorems," i.e., images produced by computer via mathematical algorithms, and perceived by the eye as integral objects, not requiring analytic proof or treatment, but having an equal integrity. At that time this suggestion would have been considered countercultural, dotty, or even heretical by the mathematical establishment. I suggested that visuals could have not only heuristic or pedagogic value, but could act as proofs having as much incontrovertibility as classical analytic proofs. If this extension of the meaning of mathematics were accepted widely, it would represent a revolution at the meta-level.

Things changed substantially in the last thirty years of the 20th century, and it is now recognized that the computer and eye combination can produce significant material that analytics often cannot reach and has no need to reach. The visual has stimulated traditional analytic methodologies that have blossomed into new chapters of mathematics, such as discrete dynamics. The "marigold" figure at right, generated in spiral fashion by the stated iteration, was produced by computer graphics but might have been obtained in pre-computer days. At that time, the interest in such things was low. Now the eye picks out many features that are of mathematical interest and for which a symbolic proof might be demanded. For example, in the concentric rings of "petals," how many petals are there in the

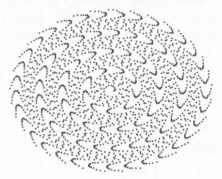

The Marigold, generated by the following iteration:
$$a = \exp(\pi i/4); \; b = \mathrm{conj}(a);$$
$$z_1 = 1; \; z_{k+1} = az_k + bz_k/\mathrm{abs}(z_k)$$

nth ring? How many dots are there in each petal? How rapidly does the figure expand from the center? The visual marigold can suggest innumerable such features, whose symbolic proofs may be trivial or perhaps exceedingly difficult. But the marigold has a visual *gestalt* whose totality cannot be described in a few mathematical sentences. To reduce the figure to mere heuristics is to degrade the visual.

Nevertheless, an attitude (or prejudice) still exists that rigorous mathematics is to be found only in analytics. Thus, speaking of recent developments in topology, Thomas Banchoff has written the following to me:

> Computer graphics showed up to save some of us. A number of high-ranking mathematicians, notably Dennis Sullivan and William Thurston, started using visualization seriously in their work. This turnaround persists. Another breakthrough was in minimal surfaces and constant-mean curvature surfaces where the computer turned out to be an essential tool in discovering properties of key examples that led ultimately to rigorous analytic proofs and a new burst of activity in a long-dead field.

Michael Emmer writes

> One of the most interesting examples of using computer graphics is the proof by David Hoffman, William Meeks, [and] J. T. Hoffman of the Costa surface as a solution of an old conjecture on infinite minimal surfaces that are not self-intersecting with topological genus greater than zero.

Thurston's Hyperbolic Knotted Wye III, from Helaman Ferguson's *Mathematics in Stone and Bronze,* Meridian Creative Group, 1994

Computer graphics, with its significant mathematical underlay, has affected both static art and animation. Peter Weibel thinks that graphical representation in science is becoming triumphant and that art—in the classical sense—is becoming obsolete.

> With the advent of the fractal, we experienced a triumphant return of the image to mathematical sciences. From mathematics to medicine, from computer-supported proof methods to computer tomography, we see an iconophilic science trusting the representative power of the image. We therefore live in a period where art, as the former monopolist of the representative image, has abandoned this representative obligation. Even all media theory is critical of the role technical images in art and entertainment. Yet science, in contrast, fully embraces the options which technical machine-based images offer for the representation of reality. Through science, the image is developed one step further, in a useful way.

There it could be the case that mankind will find images of science more necessary than the images of art. Art is threatened with becoming obsolete because of its obsolete image ideology, and it is threatened with being marginalized if it does not try to compete with the new pivotal role of the image in the sciences by developing new strategies of image-making and visual representation. Art must look for a position beyond the crisis of representation and beyond the image wars, to counterpoint science.

Today there are centers for geometry and visual mathematics and there is at last one journal, *Communications in Visual Mathematics*. A final word: Moderation, ecumenism, the attitude of "both/and" as opposed to "either this or that" is a conservative view, and though it may be the most reasonable and the most productive, it rarely raises a spirit. While there is a general—occasionally reluctant—acknowledgment that visuals can go beyond pedagogy and heuristics, the spirit of "both/and" needs to be fostered. It may be that the important question is not whether visuals lead to new results in traditional symbolically presented mathematics, but whether visuals lead to a corpus of significant mathematical experiences that carry their own semiotic content and may or may not be connected with traditional material.

Mathematical semioticist Kay O'Halloran has written, "We need literacy in all sense of the word: linguistic, visual, and symbolic. Mathematics is a multisemiotic enterprise."

Let me also mention here the extraordinary accomplishments of geometry in the area of computer animation and movies. The commercial and popular releases of such companies as Pixar have a substantial geometrical underlay that embraces approximation theory, methods of numerical analysis, differential geometry, and eigenvector analysis. The days when I used to teach the theory of cubic spline curves by bringing into class a flexible saw-blade tensed between a series of nails seem to be closer to the paleolithic mural paintings in the cave of Lascaux than they are to today's computer graphics.

Further Reading

For an elementary calculus text that is absolutely free of figures, see *Differential und Integralrechnung* written by a supreme rigorist, Edmund Landau. English translation available.

To pursue the conflicts and cooperation between those mathematicians, physicists, astro- and microphysicists who work with images and those who work with symbolic theories, I recommend a splendid article by Peter Galison, in *Iconoclash,* cited below. Galison begins with Poincaré, but actually, the story of the mathematical "iconoclash" began a century before Poincaré.

G. Allwein and J. Barweis, eds. *Logical Reasoning With Diagrams.* Oxford University Press, 1996.

David Corfield. *Towards a Philosophy of Real Mathematics.* Cambridge University Press, 2003.

Philip J. Davis. "Visual Geometry, Computer Graphics and Theorem of Perceived Type." *Proceedings of Symposia in Applied Mathematics*, Vol. 20, American Mathematical Society, 1974.

Philip J. Davis and James A. Anderson. "Nonanalytic Aspects of Mathematics and the Implication for Research and Education." *SIAM Review*, Vol. 21, No. 1, 1979.

Philip J. Davis. "Visual Theorems." *Educational Studies in Mathematics*, Vol. 24, 1993.

Philip J. Davis. *Spirals: From Theodorus to Chaos.* A K Peters, 1993.

Philip J. Davis. "The Rise, Fall, and Possible Transfiguration of Triangle Geometry: A Mini-History." *American Mathematical Monthly*, Vol. 102, No. 3, 1995.

John W. Dawson, Jr. *Why Do Mathematicians Reprove Theorems?* To appear.

Keith Devlin. *The Millennium Problems.* Basic Books, 2002.

Michele Emmer, ed. *The Visual Mind: Art and Mathematics.* MIT Press, 1993.

Michele Emmer and Mirella Manaresi, eds. *Mathematics, Art, Technology and Cinema.* Springer-Verlag, 2003.

Giovanni Ferraro. "Analytical Symbols and Geometrical Figures in Eighteen-Century Calculus." *Studies in History and Philosophy of Science*, Vol. 32, No. 3, 2001.

Amos Funkenstein. *Theology and the Scientific Imagination from the Middle Ages to the Seventeenth Century.* Princeton University Press, 1986.

Peter Galison. "Feynman's War: Modeling Weapons, Modeling Nature." *Studies in the History and Philosophy of Modern Physics*, Vol. 29B, No. 3, 1998.

Peter Galison. "Simulation, Imagery and the Iconoclash." In *Iconoclash: Beyond the Image Wars in Science, Religion and Art*, MIT Press, 2002.

David Gillies, ed. *Revolutions in Mathematics.* Oxford University Press, 1992.

Ivor Grattan-Guinness. *The Rainbow of Mathematics* (also appeared as *The Norton History of the Mathematical Sciences*). W. W. Norton, 2000.

Ivor Grattan-Guinness. "History or Heritage? An Important Distinction in Mathematics and in Mathematics Education." *American Mathematical Monthly*, Vol. 111, No. 1, January 2004.

Jeremy Gray. "The Nineteenth Century Revolution in Mathematical Ontology." In *Revolutions in Mathematics*, cited above.

Detlef Gronau. "Normal Solutions of Difference Equations, Euler's Functions and Spirals." *Aequationes Mathematicae*, Vol. 68, 2004.

George Hart. *Virtual Polyhedra* (http://www.georgehart.com/virtual-polyhedra/vp.html).

Gunther Kress. *Literacy in the New Media Age*. Routledge, 2003.

Joseph-Louis Lagrange. *Méchanique Analytique*. Paris, 1788. English translations available.

Edmund Landau. *Differential und Integralrechnung*. Groningen-Batavia, P. Noordhoff, 1934.

Brendan Larvor. "Proof in C17 [i.e., in the 17th Century]." *Philosophia Scientiae*, Vol. 9, 2005.

Bruno Latour and Peter Weibel, eds. *Iconoclash: Beyond the Image Wars in Science, Religion and Art*. MIT Press, 2002.

Steven J. Livesey. "Metabasis: The Interrelationships of the Sciences in Antiquity and the Middle Ages." Ph.D. Thesis, UCLA, 1984 [available on interlibrary loan].

F. W. Levi. *Presidential Address*. Address presented at Calcutta Math Society, January 1942.

T. Levin, U. Frohne, P. Weibel, eds. *Control Space: Rhetorics of Surveillance from Jeremy Bentham to Big Brother*. ZKM Center for Arts and Media and MIT Press, 2002.

Frank J. Malina, ed. *Visual Art, Mathematics and Computers: Selections from the Journal of Leonardo*. Pergamon, 1979.

Richard Mankiewicz. "The Muses of Mathematics." In *Mathematics, Art, Technology and Cinema*, cited above.

Elena Marchisotto. "The Projective Geometry of Mario Pieri." *Historia Mathematica*, August 2006.

A. I. Miller. "Visualization Lost and Regained: The Genesis of Quantum Theory in the Period 1913–1927." In *Aesthetics in Science*, Judith Wechsler, ed. MIT Press, 1978.

Kay O'Halloran. *Mathematical Discourse: Language, Symbolism, and Visual Images*. Continuum, 2005.

Michael O'Toole. *The Language of Displayed Art*. Fairleigh-Dickinson University Press, 1994.

Moritz Pasch. *Vorlesungen über neuere Geometrie.* Springer, 1926.

L. S. Penrose and R. Penrose. "Impossible Objects: A Special Type of Visual Illusion." *British Journal of Psychology*, Vol. 49, No. 1, 1958.

Zenon Pylyshyn. *Seeing and Visualizing.* MIT Press, 2003.

Siobhan Roberts. *Donald Coxeter: The Man Who Saved Geometry.* Walker & Co., 2006.

David J. Staley. *Computers, Visualization, and History: How New Technology Will Transform Our Understanding of the Past.* M. E. Sharpe, 2003.

Peter Weibel. "The End to the 'End of Art'." In *Iconoclash: Beyond the Image Wars in Science, Religion and Art.* MIT Press, 2002.

23

When Is a Problem Solved?

A poem is never finished, it is only abandoned.
— Paul Valéry

In der Beschränkung zeigt sich erst der Meister.
(Great craftsmanship reveals itself in its restraint.)
— Goethe

Introduction

I recently spent three days participating in MathPath (www.mathpath
.org), a summer math camp for very bright students ages 12–14. One day I
asked the students to pass in to me a question that was a bit conceptual or
philosophical. From the wide variety of questions, one struck me as both
profound and remarkable in that sophisticated interpretations were possible: Elizabeth Roberts, "How do we know when a problem is solved?"

My first reaction on reading this question—handed to me on a sheet of
notebook paper—was "mathematical problems are never solved." Due to
my limited stay at the camp, I didn't have the opportunity to ask the student exactly what she meant, so her question went unanswered at the time.
I told the camp faculty—all professional mathematicians—my gut reaction.
I added that my answer was not appropriate for the present age group and
hoped that the faculty would take up the question after I'd left. I also told the
faculty that the question inspired me to write an article. Here is that article.

A Bit of Philosophy

Some problems are solved. A baker knows when a loaf of bread is done;
Yogi Berra said, "It ain't over till it's over," which implies that a baseball

game ends. But when one thinks of the problems that confront humanity: personal, medical, sociological, economic, military, problems that seem never to be solved; it is easy to conclude that to be truly alive is to be perpetually racked by unsolved problems.

For instance, when should medical procedures or clinical trials for new medical procedures be terminated? This question is currently on the front pages of newspapers and is a matter of litigation that has included the confrontation of statisticians involved in jurimetrics and legislative intervention.

Thus, we are concerned here with a fundamental and generic question: How can we be sure that we have solved a problem? More than this, how can we be sure we have formulated a proper question? We can't, because problems, questions, and solutions are not static entities. On the contrary, the creation, formulation, and solutions of problems change throughout history, throughout our own lifetimes, and throughout our readings and re-readings of texts. That is to say, meaning is dynamic and ongoing, and there is no finality in the creation, formulation and solutions of problems, despite our constant efforts to create order in the world. Our ability to create changes in meaning is great, hence our problems and our solutions change. We frequently settle for provisional, "good enough" solutions, often described as "band-aid solutions."

What Might Elizabeth Have Meant?

One might think that in the case of mathematics—that supposedly clean-cut, logical, but limited intellectual area—the situation would be otherwise. One might think that when a mathematical problem arises, then, after a while (it may be a very long while), the problem gets solved. But think again.

The question "How do we know when a problem is solved?" can be considered on a variety of levels. The lay public tends to think that mathematics is an area where there is one and only one answer to a problem.

From the point of view of a schoolteacher, the teacher, relying on habits or traditions and considering the age of the pupils, knows when a pupil has solved a problem. It is a matter of common sense.

From the point of view of the individual or group that makes up problems for daily work, tests, or contests, I suppose that the act of making up the problem already implies a more or less definite idea of what the answer is. The examiner will think the problem is solved if he gets the answer he had in mind, or possibly a variant that conforms to certain unconsciously maintained criteria.

One possible answer, appropriate to students starting algebra, might be "you know you've solved the problem when you plug your solution back into the equation and it checks." The set of possible responses that lie between this simplistic response and my seemingly dismissive "mathematical problems are never solved," span the whole of mathematical methodology, history, and philosophy. Though responses to the question are implicit everywhere in the mathematical literature, I believe that the question as framed puts a slightly different slant on this material.

What did Elizabeth mean by her question? I can only guess. Perhaps she meant, "How can I tell whether my answer is correct?" Well, what methods or practices of validation are available at ages 12–14? Yes, you can plug the answer back into the original equation and see if it checks, but this kind of check is not available for most problems. For example, what and where do you plug in when asked to add a column of numbers? If you care to employ them, processes such as "casting out nines" and estimating the sum provide partial checks for addition. At the research level, "plugging back in" can have its own problems.

You can "check your work" by doing the problem over again in perhaps a simpler or more clever way and then compare solutions. You may, in some cases, put the problem, or part of it, on a computer. You can ask your friend what her answer is and compare. You can look in the back of the book and see whether you get the book's answer. If the problem is a "word problem," you can ask whether your answer makes sense in the real world; an answer of "minus seven and a half dappled cows" is evidence of an error somewhere.

Perhaps the student, having learned that $\sqrt{2}$ is irrational, will wonder whether or why $\sqrt{2} = 1.41421356237...$ constitutes an answer. From a certain point of view, $\sqrt{2}$ can never have a completed answer. Does one have to elaborate the meaning of the three dots (...) and trot out the theory of the set of real numbers to accept this as an answer?

Iterative solutions that theoretically "converge at infinity" are frequent. They must be terminated—abandoned—and an "answer" out-
putted. Each special problem may develop its own special termination
criterion, and I know at least thirteen different criteria that are employed.
It would be useful to have a full taxonomic study of such criteria, but
I am not aware of any such study. (See, however, the work of Sven
Ehrich cited in the Further Reading section below.)

Perhaps Elizabeth, having heard from the camp faculty that some
mathematical problems have taken centuries before they were resolved,
was asking me how long she should spend on a problem before aban-
doning it. (We all abandon problems.)

Mathematical Argumentation as a Mixture of Materials

Here is a final conjecture as to what might have been in Elizabeth's
mind as she asked the question. It is not a very likely conjecture, but
it expresses a feeling that I occasionally have after reading through
mathematical material. What is the source of one's confidence that the
informal, patched-together mixture of verbal argumentation, symbol
manipulation, computation, and the use of visuals, whether in the published literature or of one's own devising, all click together properly
as presented, and result in the assertion: "Yes, that certainly solves the
problem!"

Let me elaborate.

Consider the processes and techniques used in solving mathematical
problems. The *mélange* of materials involved has been well described
by mathematical semioticist Kay O'Halloran, who studies the relationship between mathematical ideas and the symbols with which these
ideas are expressed.

> Mathematical discourse succeeds through the interwoven grammars of
> language, mathematical symbolism and visual images, which means that
> shifts may be made seamlessly across these three resources. Each semi-
> otic resource has a particular contribution or function within mathemati-
> cal discourse. Language is used to introduce, contextualize, and describe
> the mathematics problem. The next step is typically the visualization of
> the problem in diagrammatic form. Finally, the problem is solved using
> mathematical symbolism through a variety of approaches which include

the recognition of patterns, the use of analogy, an examination of different cases, working backward from a solution to arrive at the original data, establishing sub-goals for complex problems, indirect reasoning in the form of proof by contradiction, mathematical induction and mathematical deduction using previously established results.

Behind the understanding of and expertise with symbols, there are cognitive capacities that act to create and glue together the mathematical discourse as set out in the theories of Stanislas Dehaene and of George Lakoff and Rafael Núñez (see Chapter 4, "Where Is Mathematical Knowledge Lodged, and Where Does It Come From?").

Just as logicians have wondered whether further axioms are necessary for mathematics, I wonder whether further mental capacities than those just listed are required to do mathematics that is more complex than simple arithmetic. I wonder whether, as mathematics progresses, as it adds new proofs and develops new theories, we achieve additional mental capacities by virtue of the work of the brilliant mathematicians of the past. I wonder also whether semantics, semiotics, and cognitive science, taken together, are adequate to explain the occurrence of the miraculous epiphany "Yes; that's it. The problem is now solved." Psychological studies and autobiographical material have not yet uncovered all the ingredients that make up the "aha" moment.

A Mathematician's Perspective

Imagine that the question "How do we know when a problem is solved?" has been put to a professional mathematician. There is no universal answer to this question. It depends on the situation at hand. The answers for validation given to typical young math students, as described above, carry into the professional domain (e.g., the product barcode digit checks that employ modular arithmetic), but there is much, much more that must be said.

At the very outset, one might ask "Does the problem, as stated, make sense, or does it need reformulation?" There are ill-posed problems, in either the technical sense or a broader sense. One might also ask—but is rarely able to ask at the outset—does the problem have a solution? From the simplest problems lacking solutions, such as "express $\sqrt{2}$ as the ratio of two integers" or "find two real numbers x and y such that

$x + y = 1$ and $xy = 1$ simultaneously," to the unsolvable problems implied by Gödel's Theorem, potential solvability can be an issue that lurks in the background. We can be faced with the paradoxical situation that the solution to a problem may be that there is no solution.

What kind of an answer will you accept as a solution? It is important to have in mind the purpose to which a presumptive solution will be put. A so-called solution may be useless in certain situations.

Example 1. The expression of the determinant of an $n \times n$ matrix in terms of $n!$ monomials is pretty useless in the world of scientific computation. One looks around for other ways and finds them.

Example 2. Most finite algorithm problems have a solution that involves enumerating all the possibilities and checking, but this brute-force strategy is seldom a satisfactory solution and is certainly not an aesthetic solution.

Example 3. A differential equation may be solved by exhibiting its solution as an integral. To a college undergraduate who has met up with integrals only in a previous semester, however, an integral is itself a problem and not a solution. An approximation to the solution of a differential equation may be exhibited as a table, a graph, a computer program, or may be built into a chip. Is such a solution good enough in a specific situation?

Example 4. If the problem is to "identify" the sequence 1, 2, 9, 15, 16, ... will you accept a "closed" formula, a recurrence relation, an asymptotic formula, a generating function, or a semi-verbal description? Do you want statistical averages or other properties? Will you try to find the sequence in the online Encyclopedia of Integer Sequences? Or will you simply say that a finite sequence of numbers can be extended to an infinite sequence in an unlimited number of ways and chuck the problem out the window as ill-formulated? How would you even elaborate explicitly the verb "identify" so as not to chuck the problem?

Although a problem may have been solved in a particular way, the manner of solution may suggest that it would be very nice to have an alternate solution. An interesting instance of this is the prime number theorem. Originally proved via complex variable methods, Norbert Wiener (and others) asked for a real variable proof. Since the statement of the prime number theorem involves only real numbers, the demand for such a proof was possibly a matter of mathematical aesthetics. A real variable proof was given by Paul Erdős and Atle Selberg in 1949, partly independently.

Is such and such really a solution? There are constructive solutions but, as already observed, a solution may be "constructive" in principle but in practice the construction would take too long to be of any actual use, the dimensional effect or the $n!$ effect.

There are existential solutions, in which the generic statement is "There exists a number, a function, a whatever, such that..." Mathematician Paul Gordan (1837–1912), when confronted with Hilbert's existential (i.e., non-constructive) proof of the existence of a finite rational integral basis for binary invariants, asked "Is this mathematics or theology?"

Example 5. The Mean Value Theorem asserts that given a function $f(x)$, continuous on $[a, b]$ and differentiable on (a, b) there exists a ξ in (a, b) such that $f(b) - f(a) = f'(\xi)(b - a)$. Some students find this statement hard to take when they first meet up with it. The ξ appears mysterious.

Example 6. The famous pigeonhole principle: Given m boxes and n objects in the boxes, where n is greater than m. Then there exists at least one box that contains more than one object. Who can deny this? On this basis, together with some tonsorial data, one can conclude that there are two men in Manhattan that have the same number of hairs on their head. Now find the men. We have been assured that they can surely be located in a Platonic universe of mathematical mortals.

Example 7. There exist irrational numbers x and y such that x^y is rational.

Proof: Set $r = \sqrt{2}^{\sqrt{2}}$. Now, if r is rational then, since $\sqrt{2}$ is irrational, the selection $x = y = \sqrt{2}$ works. On the other hand, if r is irrational, then set $x = r$ and $y = \sqrt{2}$. Since $x^y = (\sqrt{2}^{\sqrt{2}})^{\sqrt{2}} = (\sqrt{2})^2 = 2$, this selection works.

One may ask: Is this really a solution if we can't decide whether r is or is not rational?

There are "probabilistic solutions" to problems as in, for example, the Rabin–Miller probabilistic test for the primality of a large integer. Then there are "weak" solutions. In 1934, Jean Leray proved that there is a weak solution to the incompressible Navier-Stokes equations. Is there only one such? The question appears to be still open. What, in a few sentences, is a weak solution? If there is ambiguity about the very notion of a solution, this is especially the case for a weak solution. Technically, if L is a differential operator, and if $u = f$ satisfies the equation $Lu = g$, then f is the solution. If so, then for all "test functions" $\emptyset$, $(Lf, \emptyset) = (g, \emptyset)$, where $(,)$ designates an inner product. If only the latter is true, however, f is said to be a weak solution.

Since some problems are very difficult, and even unreachable, with current mathematical theory and techniques, the notion of a weak problem, possessing weak solutions, has been introduced as a framework that allows existing mathematical tools to solve them. A strong solution is a weak one but often a weak solution is not a strong one, (technical notions from the theory of differential equations) and the relation between the two notions is still the subject of intense research. In a numerical problem, is a weak solution really a solution if it is not computable? Despite this limitation, the knowledge that a weak solution exists can have a have considerable impact.

Apparently, the meaning of the word "solution" can be stretched quite a bit. The elastic quality of mathematical terms or definitions is remarkable, and is often achieved through context enlargement.

There are cases where a problem has been turned into its opposite. Thus, the search for the dependence of Euclid's Fifth Axiom (the parallel axiom) on the other axioms resulted in the unanticipated knowledge of its independence. The Axiom of Choice was hopefully derivable from the other axioms of set theory. It is now known to be independent of them. An instruction to prove, disprove, or prove that neither proof nor disproof is possible, is a legitimate, though a psychologically unpleasant, formulation of a problem.

There are cases where a problem was believed to be solved and then later was felt to be open, not because an error was found but because there was a shift in the (unconscious) interpretation of what had been given. For this, read Imre Lakatos' classic discussion of the history of the Euler-Poincaré theorem. A very early version reads $V - E + F = 2$, where V, E, and F are respectively the number of vertices, edges, and faces of a polyhedron. But just what kind of a three-dimensional object is a polyhedron, and what are its vertices, edges, and faces? Lakatos' discussion chronicles the ensuing tug-of-war between hypotheses and conclusions so as to maintain a semblance of the original conclusion. This is known in philosophy as "saving the phenomenon."

When Is a Proof Completed?

If the problem is to find a proof (or a disproof) of a conjecture, how does one know that a purported proof is correct? Gallons and gallons of ink have been expended on this question as formulated generally. Are proofs stable over time? A half-century after D'Alembert gave a proof of the Fundamental Theorem of Algebra, Gauss criticized it. A century after Gauss' first proof (he gave four), Alexander Ostrowski criticized it. Is a proof legitimate if it is hundreds of pages long and would tire most of its human checkers? Is a proof by computer legitimate? The mathematical community itself is split over the philosophical implications of the answers to these and a myriad of similar questions.

For one criterion as to when a solution is a solution, let's go, as bank robber Willie Sutton did, to where the money is. A recent answer to this question was formulated by the Clay Mathematics Institute, which offers prizes of a million dollars for the solution of each of seven famous problems. The Clay criteria for determining whether a problem is solved are as follows:

1. The solution must be published in a refereed journal.
2. A wait of two years must ensue, after which time if the solution is still "generally acceptable" to the mathematical community, then
3. The Clay Institute will appoint its own committee to verify the solution.

In short, a solution is accepted as such if a group of qualified experts in the field agree that it's a solution. This comes close to an assertion of the socially constructive nature of mathematics. The remarkable thing is the social phenomenon of (almost) universal, but not necessarily rapid, agreement, which has been cited as strengthening mathematical

Platonism. For more on this point read Paul Ernest's book, cited in the Further Reading section below.

In applied mathematics—and I include here both physical and social models—other answers to the basic question of this essay can be put forward. Proofs may not be of importance. The formulation of adequate mathematical models and adequate computer algorithms may be all important. What may be sought is not a solution but a "good-enough solution."

In introductions to applied mathematical and philosophical texts loops are often displayed to outline and conceptualize the process. The loops indicate a flow from

(a) the real world problem to

(b) the formulation of a mathematical model, to

(c) the theoretical consequences of the model, to

(d) the computer algorithm or code, to

(e) the computer output, to

(f) the comparison between output and experiment,

then back to any one of Steps (b) to (f) at any stage, and even, perhaps, back to (a) for, in the intervening time, the real-world problem may have changed, may have been reconceived, or may even have been abandoned.

In looking over these steps, it occurred to me that one additional step is frequently missing from this standardized list. It is that (f) can lead to

(g) an action taken in the real world and the responses of the real world to this action.

This omission might be explained as follows: At every stage of the process one must certainly simplify—but not too much, else verisimilitude will be lost. The responses of the real world are both of a physical and of a human nature, and the latter is notoriously difficult to handle via mathematical modeling. Hence there is a temptation to put a diagrammatic wall around (b) to (e) to emphasize the mathematical portion, as though mathematics is done in a sanitized world of idealized concepts that does not relate to humans. Step (g) is often conflated with Step (f) and let go at that. We are living in a thoroughly mathematized world;

additional mathematizations are inserted every day by fiat and now impact our lives in myriads of ways. It is consequently vital to distinguish (g) and to emphasize it as a separate stage of the process.

What cannot be known in advance is how often these loops must be traversed before one says the problem has been adequately solved. Common sense, experience, the support of the larger community in terms of encouragement and funding may all be involved arriving at a judgment. And yet, one may still wonder whether Steps (a)–(g) provide an accurate description of the methodology of applied mathematics.

Some Historical Perspectives

One can throw historical light on the question of when a problem is solved. There are several ways of writing the history of mathematics. I'll call them the horizontal and the vertical ways. In horizontal history, one tries to tell all that was going on in, say, the period 400–300 BC or between 1801 and 1855. In vertical history, one selects a specific theme or mathematical seed and shows how, from our contemporary perspective, it has blossomed over time.

As a piece of vertical mini-history, consider the quadratic algebraic equations first met in high school. Such equations were "solved" by the Babylonians 4000 years ago. Since then, immense new problems have come out of these equations in a variety of ways: higher-order algebraic equations, the real number system as we now know it, complex numbers and algebraic geometries; group and field theory, modern number theory, numerical analysis.

Solving a polynomial algebraic equation of degree n once meant finding a positive rational solution. Today it means finding all solutions, real or complex, together with their multiplicities, and finding it either in closed form (rare) or by means of a convergent algorithm whose rate of convergence can be specified. But the generalizations of quadratic equations go further. A formal equation can be interpreted as a matrix or even as an operator equation in various abstract spaces. The equation $x^2 = 0$ trivially has only $x = 0$ as its solution when x is either real or complex. But this is not the case if x is interpreted as an $n \times n$ matrix: The nilpotent matrices solve this equation. And if you have the temerity to ask for all nilpotent operators in abstract spaces, you have raised a question without a foreseeable end.

A more recent example, of which there are multitudes: In 1959, I. Gelfand asked for the index of systems of linear elliptic differential equations on compact manifolds without boundary. The problem was solved in 1963 by Michael Atiyah and Isadore Singer, and this opened up new ramifications with surprising features, including Alain Connes' work on non-commutative geometry.

In the historical context, mathematical problems are never solved. Material, well-established, is gone over and over again. New proofs, often simplified, are produced; contexts are varied, enlarged, united, and generalized. Remarkable connections are found. Repetition, reexamination are parts of the practice of mathematics.

A Dialogue on When a Theory Is Complete

Several weeks after having written the above sections, I received a message from Stephen Maurer, one of the MathPath faculty, that several years ago he had a web discussion with one of his most philosophical students, a first-year student in Maurer's honors linear algebra class. This discussion expands the question of when is a problem solved to when is a theory complete? I reproduce here the discussion as Maurer sent it to me.

Andy Drucker: This question has been haunting me, and I know I shouldn't expect definite answers. How do mathematicians know when a theory is more or less done? Is it when they've reached a systematic classification theorem or a computational method for the objects they were looking for? Do they typically begin with ambitions as to the capabilities they'd like to achieve? I suppose there's nuanced interaction here, for instance, in seeking theoretical comprehension of vector spaces we find that these spaces can be characterized by possibly finite "basis" sets. Does this lead us to want to construct algorithmically these new ensembles whose existence we weren't aware of to begin with? Or, pessimistically, do the results just start petering out, either because the "interesting" ones are exhausted or because as we push out into theorem-space it becomes too wild and woolly to reward our efforts? Are there more compelling things to discover about vector spaces in general, or do we need to start scrutinizing specific vector spaces for neat quirks—or introduce additional structure into our axioms (or definitions): dot products, angles, magnitudes, etc.?

Also, how strong or detailed is the typical mathematician's sense of the openness or settledness of the various theories? And is there an alternative hypothesis I'm missing?

Stephen Maurer: This is an absolutely wonderful question—how do mathematicians know when a theory is done—and you are right that there is no definitive answer. The two answers you gave are both correct, and I can think of a third one.

Your two answers were (1) we know it's done when the questions people set out to answer are answered, and (2) we know it's done when new results dry up. My third answer is (3) we don't know when it's done.

An individual probably feels done with a theory when the questions that led him/her to the subject are answered (answered in a way that he feels gives a real understanding) and he either sees no further interesting follow-up questions or can't make progress on the ones he sees. Mathematicians as a group probably feel it's done when progress peters out—the subject is no longer hot and it is easier to make a reputation in some other field that is opening up. (You called this attitude pessimistic, and I'm not so keen about it either, but it shows that math, like other subjects, is influenced by more than pure thought, and it means that mathematicians are trying to optimize results/effort.)

But, finally, history shows that fields are rarely ever done. Much later a new way of looking at an old field may arise, and then it's a new ball game. Geometry is an example. The study of n dimensions was around long before vectors and dot products (there are books of n-dimensional theorems proved by classical Euclidean methods), but the creation of these vector ideas in physics led to a new blossoming of geometry.

Another example is the field of matroids, in which I got my Ph.D. Matroids have been described as "linear algebra without the algebra." Concepts such as basis and independence make sense (and have the same theorems you have seen, such as that all bases have the same size), but there is no plus or scalar multiplication! Matroids were invented in the 1930s, for a different purpose than generalizing linear algebra, and lay fallow for some time. Then, starting in the 1960s, their general value was appreciated and they sprang to life for perhaps 30 years. We might have said that we thought linear algebra was done, but since matroids are a form of linear algebra generalization, we discovered it was not done.

Now matroids are fairly quiet again; there are still papers published in the field, but the natural questions that occurred to people when the subject was fresh have been answered or people have mostly stopped trying. It has become, like linear algebra itself, a background theory that people apply when appropriate.

Reading this dialogue recalled to my mind that Felix Klein (1849–1925) and John von Neumann (1903–1957), both towering figures in the mathematical world, emphasized other sources of rejuvenation.

> Klein: It should always be required that a mathematical subject not be considered exhausted until it has become intuitively evident... (Quoted by Morris Kline.)

By Klein's criterion, and considering contemporary proofs that require hundreds of pages or are done with a computer assist, it would appear that many mathematical subjects have a long life ahead of them.

Von Neumann's answer is different and contains a cautionary message which I, as an applied mathematician, appreciate. He warned that if mathematics removes itself from its empirical sources in the "real world," it is in danger of becoming pure arts for art's sake and then split into many branches of considerable complexity but of little significance. The remedy is to return to problems suggested by the outside world.

A Possible Example of Renewal from the Outside

It would be invidious to mention a specific example of exhaustion of a field when there are people working very happily in it. But the following example and opinion is in the open literature. Classical mathematical logic, which proceeds from Aristotle through Frege, Russell and Whitehead, Tarski, and others, has lost its connection to reality and has produced mathematical monsters. The change that is suggested is to develop logics that build in theories of probability. There currently exist a number of probabilistic logics, but they are not entirely successful. Some have even suggested that we construct logics that build in "intent" in the sense of the mathematical phenomenological philosophy of Edmund Husserl.

Implications for Mathematical Education

What are some of the pedagogic implications of the discussions of this essay? Normally, the average student thinks of a mathematical problem as something for which one arrives at a single answer as quickly as possible and then moves on to the next assigned problem. Brighter students—those who will go further with mathematics—should be encouraged to think of a problem as never really finished. Other ways of looking at the problem may emerge and yield new insights. It is also important to examine a problem in relation to other parts of mathematics as well as to the historical and cultural flow of ideas in which it is embedded. Thesis students are often advised to take a "solved" problem and think about it in their own way.

My wrap up: Discovering a sense in which a solved problem is still not completely solved but leads to new and profound challenges, is one important direction that mathematical research takes. To be fully alive in the world of mathematics is to be constantly aware of this possibility.

Further Reading

David Berlinski. *The Advent of the Algorithm*. Harcourt, 1999.

Chandler Davis. "Criticisms of the Usual Rationale for Validity in Mathematics." In *Physicalism in Mathematics*, A. D. Irvine, ed. Kluwer, 1990.

Philip J. Davis. "When Mathematics Says No." In *No Way*, Philip J. Davis and David Park, eds. W. H. Freeman, 1987.

John W. Dawson, Jr. *Why Do Mathematicians Reprove Theorems?* To appear.

Sven Ehrich. "Stopping Functionals for Gaussian Quadrature." *Journal of Computational and Applied Mathematics*, Vol. 127, 2001.

Paul Ernest. "Social Constructivism as a Philosophy of Mathematics." Thesis, SUNY Albany, 1998.

Michael Finkelstein and Bruce Levin. "Stopping Rules in Clinical Trials." *Chance*, Vol. 17, Fall 2004.

Walter Gautschi. "How and How Not to Check Gaussian Quadrature Formulae." *BIT*, Vol. 23, 1983.

Ivor Grattan-Guinness. "History or Heritage? An Important Distinction in Mathematics and in Mathematics Education." *American Mathematical Monthly*, Vol. 111, No. 1, January 2004.

Morris Kline. *Mathematical Thought from Ancient to Modern Times*. Oxford University Press, 1972.

Imre Lakatos. *Proofs and Refutations*. Cambridge University Press, 1976.

George Lakoff and Rafael Núñez. *Where Does Mathematics Come From: How the Embodied Mind Brings Mathematics into Being*. Basic Books, 2000.

David Mumford. "The Dawning of the Age of Stochasticity." *Rendiconti. Mat. Accademia dei Lincei*, tome 9, 2000.

John von Neumann. "The Mathematician." In *The Works of the Mind,* Robert B. Heywood, ed. University of Chicago Press, 1947.

Kay L. O'Halloran. *Mathematical Discourse: Language, Symbolism, and Visual Images*. Continuum, 2005.

M. O. Rabin. "Probabilistic Algorithm for Testing Primality." *Journal of Number Theory*, Vol. 12, 1980.

Megan Staples, e-correspondence.

Herbert Wilf. "What Is an Answer?" *American Mathematical Monthly*, Vol. 89, 1982.

For the Clay Institute criteria, see the following website:
http://www.claymath.org/millennium/Rules_etc/

What Is Meant by
the Word "Random"?

A random sequence is a vague notion ... in which each term is unpredictable to the uninitiated and whose digits pass a certain number of tests traditional with statisticians.

—D. H. Lehmer (number theorist), 1951

The extensive theorizing and varying interpretations of probability are children of humanity's continuing love affair with gambling. Mathematizations arrived late in on the scientific stage: the 16th century. The theory of probability now inextricably tied to mathematical statistics, which came a century later, and an indispensable tool in science, technology, and social and economic studies, is a rich source of conflicts with intuition and common sense. The paradoxes of probability theory have filled numerous books, and Benjamin Disraeli's gag that there are "lies, damn lies, and statistics" is trotted out frequently. This essay and the next two offer a few observations relating to the conflict.

Probability theory is founded on the notion of the "random event." But are there random events in the world? While in a deterministic frame of mind, the poet Alexander Pope wrote "All nature is but art, unknown to thee; All chance, direction, which thou canst not see."

Albert Einstein agreed with this and said no, God doesn't play dice with the universe. Niels Bohr snapped back: "Albert, don't tell God what to do."[1] Quantum theory in physics is probability-laden. Is

---

1. On the other hand, Stephen Hawking is supposed to have remarked that God played dice only once. This has been explained to me as follows: at the beginning of the creation, as the result of the breaking of certain symmetries as the universe began to cool, the values of the fundamental numbers were established. After that, God had no choice but to allow that universe to go forward.

causality an outworn concept and free will reestablished? What does it mean when the stage magician says "Pick a card at random; any card"? What does it mean when a pollster says "A random sample of voters shows that..."?

But what does "random" mean? Chaotic? Irregular? Thoroughly mixed up? Unpatterned? Featureless? Chancy? Accidental? Totally disordered? Lawless? Unformalizable? Unpredictable? Ununderstandable? Are there degrees of randomness? Can a single isolated event be said to be random? What constitutes an isolated or independent event? Can the notion of a random selection or a random sequence (often called a random number stream) be formalized and computerized? Or is randomness merely a psychological conception or a state of mind?

Note also that randomness enters into ethics and into the law. Is placebo testing on random patients for a new drug ethical? Is random searching of passengers on airlines and trains unconstitutional? After all, the Fourth Amendment to the U.S. Constitution says

> The right of the people to be secure in their persons, houses, papers, and effects, against unreasonable searches and seizures, shall not be violated, and no Warrants shall issue, but upon probable cause, supported by Oath or affirmation, and particularly describing the place to be searched, and the persons or things to be seized.

We mean many different things when we say the word "random," and some appear to be inadequate or contradictory. Thus, if we say that a sequence of digits is random if each digit appears with frequency 1 in 10, a perfectly understandable requirement, then the repeated repetition of the sequence 0, 1, 2, 3, 4, 5, 6, 7, 8, 9, ... has this property yet is perfectly patterned and predictable.

If you strengthen this requirement by saying that every pair of digits (such as [3, 8]) must appear with frequency 1 in 100 and even that every conceivable n-tuple of digits must appear with frequency 1 in 10^n, you have defined what is known as *base-10 normality*. Then comes the paradoxical fact that the Champernowne sequence,

0 1 2 3 4 5 6 7 8 9 10 11 12 13 14 15 16 17 18
1 9 2 0 2 1 ...,

arrived at by writing the integers successively, is provably base-10 normal and yet is perfectly patterned and computable. On the other hand, the successive digits of π,

$$3 \ 1 \ 4 \ 1 \ 5 \ 9 \ 2 \ 6 \ 5 \ 3 \ 5 \ 8 \ 9 \ 7 \ 9 \ 3 \ 2 \ 3 \ 8 \ 4 \ ...,$$

which are relatively easily computed, have not yet been proved to be base-10 normal, although the computational evidence extracted from billions of digits of π seems to confirm the conjecture.

Mathematicians have racked their brains in an attempt to characterize randomness in a formal manner. In 1919, the Austrian-American mathematician Richard von Mises (1883–1953) defined randomness as the impossibility of deriving a successful gambling system (say, at roulette). Von Mises tightened this vague notion, a notion that reflects centuries of experience, by saying that a random sequence is such that, given any rule that selects out a subsequence, the numbers of the subsequence exhibit the appropriate frequencies.

But just what constitutes a "selection rule"? And this question moved the theorists into the notion of a rule that is computable by a digital computer or of its formalization known as the Turing Machine. Von Mises' definition was found inadequate by the community of probabilists, yet Andrei Kolmogoroff (1903–1987) wrote

> The basis for the applicability of the results of the mathematical theory of probability to real "random phenomena" must depend on some form of the frequency concept of probability, the unavoidable nature of which has been established by von Mises in a spirited manner.

Another stab at a definition has been ascribed independently to Ray Solomonoff, Andrei Kolmogoroff, and to Gregory Chaitin. It says that a string is patternless if it cannot be described or specified more compactly than the string itself.

As an example, consider the string 6 7 9 7 1 4 12 17 86 21. How would you describe it or create it with fewer symbols (bits) than the string itself has? On the other hand, the string of 18 symbols 3 5 3 5 3 5 3 5 3 5 3 5 3 5 3 5 3 5 might be described as "repeat 3 5 9 times," which requires only 14 symbols.

Another definitional candidate is that of P. Martin-Löf, which appears to be the current *beau-idéal* of definitions of randomness, but this definition is too complicated to be stated here.

Thus, there is no formal mathematical definition of a random sequence that seems to satisfy everyone's standards and everyone's needs. Yet, random number streams are used with advantage every day by scientists, and one might say that a definition of randomness is therefore a tremendously useful oxymoron.

While theoreticians have dwelt in the realms of ideal but non-computable procedures and tests, pragmatic researchers have come up with algorithms known as pseudo-random number generators. These are available in chips and in all good scientific computation packages. Using the random number command in the MATLAB package, I pulled off the following stream of one hundred numbers. Do they seem sufficiently mixed up? Does a pattern jump out and strike the naked eye?

```
6  7  9  7  1  4  9  9  4  8  0  3  8  0  1  2  1
6  2  1  0  7  4  9  4  4  8  5  2  6  8  0  6  3
8  5  7  4  3  1  1  6  3  5  1  6  3  8  8  5  4
8  8  6  8  6  3  2  3  5  7  3  8  5  3  7  5  4
6  6  7  9  5  8  1  9  2  2  8  7  1  0  8  1  2
6  2  4  0  9  5  4  5  3  4  2  5  7  5  6
```

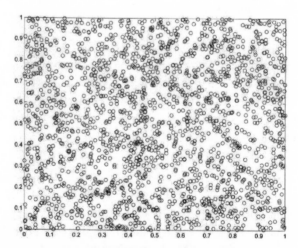

Is this distribution of points random?
Can your eye discern any patterns?

There are CD-ROMs containing billions of random bits. They have been produced by means of deterministic pseudo-random number generators (computer algorithms) combined with the digitalization of white and black noise created by physical devices.

The best schemes have had their output certified as having successfully passed a number of sophisticated statistical tests. Among the tests can be cited uniformity, poker-hand tests, gap tests, etc. Dozens of tests have been proposed.

The question of which sequences are "random enough" or which tests are necessary for a sequence to pass as random depends upon the uses to which the sequence will be put. For example, if you want to use the sequence to approximate integration, then deterministic equidistributed sequences (which are easily generated) will suffice. There is a great difference between the pseudo-random number generators used for Monte Carlo simulations and probabilistic computations of various kinds and the generators used for cryptography and computer security. The random number generators used in computer-assisted Hollywood films need to have a still different character. These distinctions are little discussed by the theorizers of ideal randomness.

The tension between an ideal and what is achievable and useful, between the conceptual infinite and the humanly finite, is striking.

Further Reading

Gregory Chaitin. *Meta Math! The Quest for Omega*. Pantheon, 2005.

Donald Gillies. *Philosophical Theories of Probability*. Routledge, 2000.

Ian Hacking. *The Taming of Chance*. Cambridge University Press, 1990.

Sérgio B. Volchan. "What Is a Random Sequence?" *American Mathematical Monthly*, January 2002.

25

The Paradox of "Hitting It Big"

A high school teacher of mathematics recently sent a letter to a local newspaper complaining that, with regard to gambling, the mathematics teachers of the country have not done their duty. They have not driven home the brute fact that, on average, people who go to gambling casinos or play the lottery lose money. I have a somewhat different opinion.

The spread of gambling is a *bête noire* of mine, and I recently passed a billboard promoting a lottery with the words "You can win multimillions. Buy a ticket." Yes, you can win, I thought; it doesn't contravene the laws of physics or of mathematics that you can win, but I've often wondered why the theorems of probability do not discourage people from laying out sums, week after week, month after month. I've wondered what mania grips the sad sacks who clog the casinos, shove bucketsful of coins into the slots, and wind up with an addiction.

People know very well that, on average, they lose money when they gamble. People know that the casino owners or the state governments are raking it in, that if they weren't, the slots and the wheels and the little scratch away slips would rapidly be junked, that the stock values of the big suppliers of mathematical gambling systems and hardware would plummet. They know (or they should know) that politicians love state-regulated big-time gambling because it provides revenue without the stigma of being called a tax. The spread of gambling, therefore, is not a question of knowledge; something else is at play.

Mary Y. and John Q. Public say to themselves: "Averages be damned! I'm not going to run my life by averages. I'm really going to hit it big next time." They may not be sufficiently mathematically sophisticated to know about averages. They have just seen the large billboard that says "You too can become a multimillionaire." And when they hit it, millions and millions of lottery dollars will walk into their bank account over a period of years, to change their lives, to make them ecstatically happy or miserable, as the case may be.

What is the relation between the mathematical theories of probability and people's actions? Is the former an adequate description of the latter? Is it possible to set a dollar value on the entertainment implicit in gambling, on the expectation of "hitting it big," and to factor the adrenaline into our theories? (Some psychologists have written that with numerous people, the adrenaline actually comes from losing and not from winning.) Well, you might say that the market automatically does the factoring in, and its answer seems to have resulted in the greatest mania for gambling and the production of the greatest legalized, mathematized, computerized gambling industry that has ever flourished in this country.

Moving away from gambling conceived narrowly as the Lot, the Wheel, or the chances that kids sell to enable the local high school band to buy uniforms, we are all gamblers. When we cross the street we take chances; when we beget children we take genetic chances. The pop-philosophers even assure us that life is a worthwhile gamble, so we indoctrinate our children into our personal perception of the common sense aspects of living in a chancy world. And what role does mathematics play in all this?

Some years ago I went to a scientific meeting held at the National Cancer Institute in Bethesda, Maryland. A noticeable number of the participants, including employees of the Institute, were puffing away on cigarettes during the breaks between talks. How would it be now that the antismoking campaign has intensified considerably? What changes of habits have resulted from bumper stickers that read (generically) "Say NO to X, Y, or Z"? Why is the air in American restaurants smoke-free, while in some foreign countries the opposite is frequently the case? Don't they read the stats?

Consider the gambling aspects of book publishing. There are many reasons why a person might want to have a book published: the desire to inform, to amuse, to warn, to confess; the desire to appear wise, to achieve fame or notoriety, to praise, to smear, to amass power; the desire to perform a public service. And there is one additional motive, lurking behind all the others: the desire to hit it big with royalties. We have all heard of books in the scientific field that have hit it big: text-books, popular science books, even books that only a half-dozen people in the world could claim to have understood. Out of the hundreds and hundreds of manuscripts submitted in these categories, many have been called, but Lady Luck has chosen only a few to hit it big.

I have not seen any statistics on the subject, but my personal esti-mate is that, on average and on an hourly basis, book-writing pays less than minimum wage. You have written, say, a short monograph on an advanced mathematical topic. How many copies will you sell? 1500 all over the world? (You might have sold 3000 in the palmy days before libraries were computerized and sharing their holdings.) So at $40 a copy, and with a royalty of 10%, you will get $6000. If you worked on your book 300 days for 4 hours a day which is 1200 hours, your book would earn you $5 an hour, which is considerably less than the mini-mum wage in my state. Do you think you could knock off a monograph in that time? Could a novelist knock off a novel in that time? No way, José, unless the novelist's name happens to be Georges Simenon or Barbara Cartland. And I've read that the average first novel sells fewer than 2000 copies.

Of course, all kinds of objections can be raised to these computations. Nonetheless, on the basis of the statistics, I would advise a would-be author to get out of the writing game and into lecturing or op-ed-ing, where the hourly return can be greater. Yet the flame lit by the expecta-tion of "hitting it big" burns brighter than you might think. The experi-ence of J. K. Rowling, author of the Harry Potter books, is seductive. I know a number of scientists who have taken up the pen in this expecta-tion only to be gravely disappointed.

What about the view from the publishers' side? Publishers try to diversify their list of authors. What they lose on monographs, they may make up on wildly successful textbooks or popular science titles. So, we might say weep no tears for the publishers just as we say the same for casino owners.

Enough about writing. After all, relatively few people write for money. Let's talk about travel. A greater fraction of the population travels, after all. But consider this: What is the prudent course of action to take when you read in the papers that a certain airline has had a disaster and you are holding space on one of their flights? We experience the reverse phenomenon, in which a single disaster sets off a chain reaction in excess of what might have been deduced from a calm consideration of the probabilities. In response to terrorist action in London, the tourist business experienced a terrible slump. When my wife and I showed up in London as we had planned months in advance, our British friends expressed surprise: "You didn't cancel out?"

Yes, one day we say, "Damn the stats. The average weather over a lifetime is a light drizzle, but there will be sunny days and then we'll get to the beach. All honor to the individual occurrences!" And the next day we succumb to the insurance peddler who points out some corner of our lives that is not covered by the policies we now hold.

Apart from my personal insights and reactions to the activity known as gambling, there is the psychological literature, due especially to the Nobelists in economics, Amos Tversky and Daniel Kahnemann. The literature shows that people's evaluation of likelihoods is utterly inconsistent with the most basic constraints of probability theory and the most basic constraints of utility theory. Here are some quotes from a book by Ernest S. Davis:

> Many studies have shown that human subjects judge likelihoods in bizarre ways. In one experiment undergraduate subjects were presented with the following personality sketch.
>
> "Linda is 31 years old, single, outspoken, and very bright. As a student she was deeply concerned with issues of discrimination and social justice and also participated in antinuclear demonstrations."
>
> The subjects were then asked which of the following statements was more likely:
>
> (a) Linda is a bank teller.
> (b) Linda is a bank teller who is active in the feminist movement.
>
> 86% of the subjects judged the second statement more likely, despite the fact that this violates the basic rule that a more general state (being a bank teller) is never less likely than a more specific state (being a feminist bank teller). Perhaps more surprisingly, 43% of the psychology graduate students who were given the same question made the same mistake. It is hard to conceive of a coherent theory of likelihood that would justify this answer.

A second experiment displays inconsistencies of choice. Subjects are divided into two groups, I and II. Group I reads the following paragraph:

"There is a community of 600 people being threatened by an epidemic. The authorities have the choice between two policies. If they follow policy A, there is a 100% chance that they will save 200 people. If they follow policy B, there is a 1/3 chance that they will save all 600 and a 2/3 chance that they will save no one. What should they do?"

For Group II, the two alternatives are changed to the following: "If they follow policy A, there is a 100% chance that 400 people will die. If they follow policy B there is a 1/3 chance that no one will die and a 2/3 chance that everyone will die."

Subjects in Group I overwhelmingly preferred policy A while subjects in Group II overwhelmingly preferred policy B. But, of course, the two choices are the same; just that the choices from Group I are phrased positively while the choices from Group II are phrased negatively.

Nonetheless, people carry out plausible reasoning continuously, and they do it well enough to work their way through a very uncertain world. They may be doing something right, but it will not be easy to explain what they are doing and why it is inadequate.

Conclusion: It is very hard to move from mathematical theories of probability to economic or moral advice for the individual. And if such advice is given, whether or not it will be taken seems itself to be random.

Further Reading

Lorraine Daston. *Classical Probability in the Enlightenment*. Princeton University Press, 1988.

Ernest S. Davis. *Representations of Common Sense Knowledge*. Morgan Kaufmann, 1990.

Robyn M. Dawes. *Rational Choice in an Uncertain World*. Harcourt Brace Jovanovich, 1988.

Ian Hacking. *The Emergence of Probability*. Cambridge University Press, 1984.

Michael and Ellen Kaplan. *Chances Are: Adventures in Probability*. Viking, 2006.

Amos Tversky and Daniel Kahnemann. "The Framing of Decisions and the Psychology of Choice." *Science*, Vol. 211, 1981.

Daniel Kahnemann and Amos Tversky. "On the Study of Statistical Intuition." *Cognition*, Vol. 11, 1982.

26
Probability and Common Sense: A Second Look

La théorie des probabilités n'est que bon sense confiné par le calcul.
—Pierre Simon Laplace, 1796

At our current level of understanding, common sense reasoning and probability theory are linked inseparably but unhappily, like partners in a codependent but dysfunctional relationship. Probability depends on common sense reasoning in that the construction of a probabilistic model for a real-world domain generally depends on a common-sense understanding of the domain. Constructing such a model involves the following steps:

1. identifying a significant set of events,
2. positing independence assumptions, and
3. assigning numerical probabilities.

Of these three steps, statistical analysis of a data set is often, though by no means always, sufficient for the third. It is occasionally sufficient for the second. And it is of almost no value for the first.

What cannot be done from statistics must rely on a prior understanding of the domain. Common-sense reasoning is also generally needed to evaluate and reformulate data before statistical analysis can be done. In most cases, before statistical analysis can be performed on a data set, data items that are obviously corrupted or redundant may be excluded, features in the data set that will merely confuse the analysis may be excised, features that are needed for analysis but not directly expressed in the data set may be added. Generally, none of this can be done without a prior understanding of what the data means, how the data was collected, and so on.

In most applications of probabilistic reasoning, however, it is necessary to make independence assumptions that, to a common-sense understanding, are not even close to true but are wildly fantastic when taken literally. For example, the "trigram" model of natural language has been quite successfully applied to the problem of tagging each word in a text with its part of speech, such as determining whether an occurrence of the word "can" in a sentence is a modal, a noun, or a verb. The trigram model posits that how a given word should be tagged with a given part of speech depends only on the word itself and the previous two parts of speech, and, given those, is independent of all the other words and all the other parts of speech. This assumption is obviously not even close to true in actual natural language text, and it is extremely easy to come up with innumerable counterexamples. Nonetheless, the manifest falseness of this assumption somehow comes out in the wash, and the model works quite well. Moreover, improved models that make more realistic assumptions about natural language run the risk of being too complex and requiring too much data to be usable. Thus, using probability often involves deliberately going against one's own common sense.

On the other hand, common-sense reasoning depends on probability in the following way: In almost all cases, common-sense reasoning involves drawing conclusions that are uncertain and provisional from premises that are likewise uncertain. It therefore depends on the ability to deal with uncertainty and to combine uncertainties. Probability theory, in the subjective, or Bayesian, interpretation[1] is by far the most powerful, flexible, and useful calculus of uncertainty that has been found, despite a substantial body of research that has sought alternatives.

In the current state of the art, however, probability theory has been successfully applied to common-sense reasoning only for narrowly defined problems in restricted domains. It is entirely unclear how it can

---

1. The subjective, or Bayesian, interpretation of probability theory construes probability theory as being concerned with the likelihood of sentences. Alternative interpretations are the objective theory, which construes probability theory as concerned with the fraction of a sample space satisfying a given property, and the frequentist theory, which construes probability theory as concerned with the frequency with which a given experiment will have a given result.

be applied, or whether it will suffice, for common-sense reasoning in a broader sense.

Further Reading

Ernest S. Davis. *Common-Sense Reasoning and Mathematical Modeling.* In preparation.

27

Astrology as Early Applied Mathematics

Astrologer to fat client: Your astrological chart tells me you need to lose
a few pounds... Your moon has been in the House of Pancakes.
—Jeff MacNelly, *Shoe* (comic strip), February 28, 2000

I would like to think that this epigraph summarizes the average American opinion today about astrology, but I doubt that it does. The zillions of dollars that people here and around the world spend annually on astrology lead me to conclude otherwise. A fraction of this money is spent on sophisticated Web programs that, to the average person, contain deep and arcane mathematical computations. What is the common sense of astrology? It depends whom you ask, and the answer is time-dependent. If one consulted the physician and astrologer Simon Forman (1552–1611), he might very well cast a horoscope and ask for a urine sample.

The case of Girolamo Cardano shows how closely interest in mathematics was linked to the practice of astrology. Girolamo Cardano (1501–1576) is known to the community of mathematicians as the leading mathematician of his day. Though not the discoverer of the solution to the general cubic equation, he was the first to publish the formula and, as a consequence, one of the first mathematician to puzzle over the existence and nature of complex imaginary numbers. He is also known among mathematicians as one of the founding fathers of probability theory.

But Cardano was also a practicing physician, a professor, a natural philosopher, an experimental gambler, a prolific author and autobiographer, and, last but by no means least, a practicing astrologer and an interpreter of dreams. As a minor accomplishment, he was the inventor of the citation index, a contribution for which, I suppose, today's selection committees are most grateful. He was volatile and outspoken in his criticisms of standard medical practices. He was a heretic, condemned to house arrest by the Inquisition. His complete works in ten volumes, first published in Lyons in 1663, were reprinted in 1966. His autobiography, which is remarkable for its candor, is available in English and still bears reading.

What were some of Cardano's accomplishments as a practicing astrologer? He established genitures (times of birth or of other key personal events and the related stellar configurations) and interpreted them; he wrote, published, experimented (often using his own medical problems and moods); he read, criticized, and "improved" on the established astrological procedures. He collected, collated, and cast horoscopes for the well-known: Charles V, Sultan Süleyman, Francesco Il Sforza, Pope Paul III, and Jesus, among many others. He wrote a code of ethics for astrologers. (Will we shortly need one for mathematicians?) He admitted that he had his own "batting averages" and advised caution for fellow and prospective astrologers, stating that at the bottom line, the practice of astrology was (in today's lingo) part algorithm and part a personal art. He spent a lifetime fine-tuning both parts.

Thus, Cardano was a man who gave the term "Renaissance Man" its meaning. Anthony Grafton has written a most excellent and sympathetic portrait of Cardano in his persona as astrologer and of his astrologic practice, placing it within the mindset of the 16th century. "Cardano's palette glowed with many colors, not just the black and gold of astrology, but the red of medicine."

The basic premise of astrology is that the world is decipherable and that the celestial bodies and our relation to them hold the key to our personal history and to world history. The language of astrology is still embedded in common expressions: We "thank our lucky stars." We speak of jovial, mercurial, saturnine, martial personalities: Aquarians are thus and so; Sagittarians are quite different. Relating the planets to parts of the human body, to human thoughts and intentions, and to material objects such as precious stones and metals, or to the constellations and the "houses" in which they occur; considering carefully the conjunctions and oppositions, quadratures, trines, and sextiles of the planets; regarding the exact moments of a person's birth and the location of the planets at that moment—these are all constituent parts of astrologic recipes for interpretation. Cardano's work represents astrology at its most advanced, most theoretical, most computational hour, set forth as a belief and a social practice. This is astrology to which both mathematics and astronomy owe a debt because of its insistence on the production of accurate planetary tables

During all of its existence, astrology has been suspect, derided as mere superstition, and a century ago historians of science regarded it with nothing but distaste and proceeded to ignore it. Its ideas have come to be regarded as an embarrassment, particularly when its practitioners—as in the case of Cardano—also had accomplishments that are now regarded as fundamental. More than distaste and avoidance, there has been denial. Indeed, the authenticity of the *Tetrabiblos* (four books concerning the influence of the stars), written by the great Claudius Ptolemy, the most famous astronomer of antiquity, and highly regarded for fifteen hundred years, came to be questioned in the early 19th century. How was it possible for such a great scientist to have been so self-deceived? Somewhat later, through textual comparison with Ptolemy's *Almagest*, Franz Boll re-established Ptolemy as the author.

"Trash is trash," a scholar once remarked, "but the study of trash can be scholarship." What we have in Grafton's book is a shift from astrology as a discredited superstition to astrology as a social and cultural arrangement. Why the turnaround? Until the advent of philosophical relativism, astrology, like the doctrines of kabbala or the practice of talismans, were not regarded as ideas worth recreating or reanimating. These systems were random, confused charlatanry, in which no one ever really believed. The idea, now popular, that there are many possible world-views (for example, the theory of intelligent design), or that each person may have an individual world-view, has opened up these and many other topics for scholarly study. It is part of today's multiculturalism. Depending on the ideas discussed, this shift occurred at different times in the 20th century. The introduction of ethno-mathematics into discourses on mathematical education may be a late instance of such a shift.

Grafton's book raises another question in my mind: How are we now to regard astrology as practiced by someone like Cardano? Is it trash or is it science? Well, what are the hallmarks of a science: a sound metaphysical basis? observation? experimentation? reproducibility? empirical verification? the back-and-forth of fine-tuning? a theory expressible in mathematical terms and within which manipulations lead to conclusions, verifiable or not? the potential falsifiability of its statements? acceptance by a knowledgeable elite? Although a definitive definition of science is impossible, all the attributes above, in various combinations, have been proposed, and all are manifested in Cardano—including the idea that we are affected by the stars. After all, Newton's theory of gravity says that the stars and I experience a mutual (albeit tiny) attraction. All this being the case, one can argue with some cogency that astrology as practiced by Cardano was a science. A failed science, certainly, and a social practice with certain consequences, but not a pseudo-science if such a distinction is valid.

Although this judgment probably runs counter to the prejudices of some readers, let me ask: "What is applied mathematics?" This question arises when one considers a number of acrimonious interchanges that have arisen as to whether such-and-such is a legitimate application of mathematics. An acceptable answer might be: "Applied mathematics is mathematics in the service of matters that are exterior to mathematics itself." In that service, the question of logical proof or truth or finality is not really an issue. What matters is utility or significance, in some shape or form, to a sufficiently large group of people. With this criterion in mind, one can assert that Cardano's astrology constituted a non-negligible chapter in the history of applied mathematics. One may also wonder how future generations will regard our own applications of mathematics.

Further Reading

Anthony Grafton. *Cardano's Cosmos: The Worlds and Works of a Renaissance Astrologer*. Harvard University Press, 1999.

Oystein Ore. *Cardano, the Gambling Scholar*. Princeton University Press, 1951. Contains a translation from the Latin of Cardano's book on games of chance, by Sydney Henry Gould.

Barbara Howard Traister. *The Notorious Astrological Physician of London: Works and Days of Simon Forman*. University of Chicago Press, 2001.

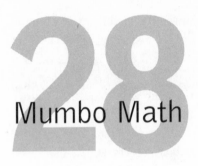

Mumbo Math

I once undertook to review Francis Wheen's *How Mumbo-Jumbo Conquered the World: A Short History of Modern Delusions*, under the mistaken impression that it contained a section on the current mania for gambling. I wanted to see what Wheen, one of the top-notch debunkers currently writing, had to say about the gambling situation, but I found that his book had other things to say. While *Mumbo-Jumbo* concentrates on economics, business, politics, international relations, dire predictions, dianolatry (remember Princess Diana?) and the empty pretentious rhetoric of some academics, there are a few pages that in some way relate to mathematics.

Francis Wheen, author, journalist, "Columnist of the Year" for the (UK) *Guardian*, skeptic, satirist, wit, liberal, radio and TV personality, has for years described and exploded the follies, the delusions, the manias, the lies, the scams, that overtake humankind. You might think that a long litany of current stupidities would be depressing, but not at all. Wheen's satiric pen makes this book delightful to read, and I recommend it to one and all. At the bottom line, Wheen's message is that the Age of Enlightenment, the age that promoted rational investigation and argument as the road to truth and the foundation of objective knowledge, is being supplanted by an Age of the Irrational, the Mystic, and the Sensational.

Let's first consider the familiar pigeon's nest of ideas that embraces the number 666 (the "number of the beast" in the *New Testament's Book of Revelations*), the sayings of the apocalyptic catastrophists, Y2K (the largely unwarranted angst that accompanied entrance into the new millennium), the Bible Code, spiritualism, and that sort of thing. Did you know that the sum of the squares of the first seven primes totes up to 666? Did you know that the fourth root of π is 1.331 335 363... and that $331 + 335 = 666$?

If these facts thrill you, if they send a frisson up your spine, then you are a bit of a numerologist, a neo-Pythagorean. If you attribute arcane or transcendental potency to individual numbers; if you read into these relations more than mere arithmetic coincidences, then you, as well as thousands of others, have a numerological gene. Not to worry. You are in good company: A number of modern cosmologists have such genes, and so did John Napier. If you look for secret messages in the Bible by taking every other letter, you are a gematria freak and are in a plentiful company that once included Mozart.

Consider next a pronouncement of feminist Luce Irigaray. She denounced Einstein's $E = mc^2$ as a "sexed equation since it privileges the speed of light over [less masculine] speeds that are vitally necessary to us. Whereas men have sex organs that protrude and become rigid, women have openings that leak menstrual blood and vaginal fluids." Then she goes on to imply that the problems of turbulence in fluid theory cannot be solved with the traditional masculine mathematical methodologies. Admittedly—as G. I. Barenblatt pointed out—very little pure theory of turbulence has emerged since the days of Ludwig Prandtl, Theodore von Karman, Geoffrey I. Taylor, Andrei Kolmogoroff, et al., but I think I know how the ghosts of these applied mathematicians would react to Irigaray's protest.

Similarly, consider a statement of Jacques Lacan, a psychoanalyst and one of the high priests of structuralism.

> Thus the erectile organ comes to symbolize the place of *jouissance* [ecstasy], not in itself, or even in the form of an image, but as a part lacking in the desired image: That is why it is equivalent to the $\sqrt{-1}$ of the signification produced above.

Wheen calls this the "tyranny of twaddle" and I agree. Without defending Lacan's use of the imaginary unit in this instance, I admit that we mathematicians do not have patent rights on the symbol $\sqrt{-1}$. It's in the public domain, and the variety of uses to which it is put become dissertations of semioticians. I've noticed that certain mathematical terms have in recent years gone public. Consider the word "matrix." Formerly the property of anatomists, geologists, and printers, it entered mathematics in the 1800s. In my parochialism I consider this noun to be so mathematical that I attribute its current prominence in products, from hair styles to movies to cars the result of students who took Math 106 and upon graduation found themselves on Madison Avenue. I imagine that they imported this term to add solemnity, scientific verisimilitude, and mystery to the products they hawk.

Texts such as Irigaray and Lacan are wide open for parody. I suppose that every academic, every scientist has now heard of the article *Transgressing the Boundaries: Toward a Transformative Hermeneutics of Quantum Gravity*, a spoof, or parody, that physicist Alan Sokal wrote and got published in a professional journal (1996), much to the chagrin of the journal's editors.

Wheen has this to say:

> Sokal's article was littered with scientific howlers and absurdities. He claimed that Jacques Lacan's Freudian speculations had now been proved by quantum theory, and that Jacques Derrida's [philosopher of deconstructionism] thoughts on variability confirmed Einstein's general theory of relativity.

One sentence of Sokal's spoof will suffice for my purposes here:

> In mathematical terms, Derrida's observation relates to the invariance of the Einstein field equations $G_{\mu\nu} = 8nT_{\mu\nu}G_{\mu\nu} = 8nGT_{\mu\nu}$ under nonlinear space–time diffeomorphisms.

Who in his right mind would fall for "such tosh as this?" Wheen points out a twist that is relevant to the Sokal story. Recently, Andrew Bulhak, of Monash University in Australia, produced a "post-modernism generator" that produces "random, meaningless and yet quite realistic text in genres defined using recursive transition networks." You can find a sample document at www.elsewhere.org/cgi-bin/postmodern/, and you will find directions there for generating additional documents of your own. Actually, this kind of thing has been done for at least twenty-five years and is easy to do, although perhaps not so easy in the mathematical field, where, according to a computer scientist correspondent of mine,

> It would be difficult to do so that it was funny. For the joke to work, it has obviously to be nonsense. Mathematical nonsense is too hard to tell from mathematical sense.

When I first heard of the Sokal "scandal," I thought that Sokal's point was to unmask the absurdity of pretentious pseudo-mathematical invasions of humanistic fields. I said to myself: "It's about time that someone did it."

On the other hand, I have learned much to my embarrassment that there is a certain danger in writing spoofs. Many (pre-terminal) years ago, as a reaction to the rapidly multiplying class-action suits of the time, I wrote a spoof in which I said that a certain left-handed person instituted such a suit in the New York courts on behalf of the lefties of the country, claiming discrimination in that they had to drive on the right side of the road. I laid my narrative on quite thickly and put in a few legalistic phrases and pointed out, of course, that the question of left–right orientations attracts a bit of attention in mathematics and physics.

My spoof delighted the editor of a national magazine and he ran it. The article—or at least its title—then got placed in some database (yes, there were searchable databases before Google and others got into the act), and a few years after the article appeared, I got a phone call from a young researcher at the one of the TV networks. She told me she was working on a show relating to the problems of the disabled and wanted to have more of the legal details of the suit that I had described. After a moment's hesitation ("Oh, what a tangled web we weave, When first we practice to deceive!"—Sir Walter Scott), I broke down and confessed the truth of the matter. Moral: If you write a spoof, no matter how preposterous it is, there's always someone out there who will believe it.

"What fools these mortals be," observed Puck in *A Midsummer's Night Dream*. Reader, can you imagine what it would be like to live in a world free of bunkum, scams, stupidities, rigidities, misconceptions, and misrepresentations? when applied mathematics is free of folly? Measure the heartbreak it would save against the laughter that would be exiled.

Further Reading

Philip J. Davis. "Paul Feyerabend: Philosopher or Crank?" *SIAM News*, December, 1995.

Niels Møller Nielsen. "Counter Argument: In Defense of Common Sense." Dissertation, Department of Language and Culture, Roskilde University, Denmark, 2000.

Francis Wheen. *How Mumbo-Jumbo Conquered the World: A Short History of Modern Delusions.* Fourth Estate, 2004.

29 Math Mixes It Up with Baseball

I suppose we have all heard of Franglais, the mixture of English and French that is increasingly frequent in spoken French and which irritates French language purists. There is also Angleutsch, the invasion of English terms into German. I suppose that because English is, at the moment, the most popular language for intercommunication, every language has a corresponding mix.

What amuses me is the invasion of mathematical terms into spoken or written English. On a recent political discussion on TV, I heard a panelist say that "policy A was the square or the cube of policy B," in the sense that its provisions were much stronger. The mathematical purist, criticizing the TV commentator, might point out that the square or the cube of x is greater than x only if x is itself greater than one (e.g., $(1/2)^2 = 1/4$). Then there are the oxymoronic statements: All people in Glendale Heights are above average. (OK, then, who is average?)

The word "solutions" is very common as the name of companies: Hair Solutions (not a rinse, just a beauty salon), Investment Solutions. Consider also: *Vector* Shoes, *Matrix* Enterprises, life's *spiral* path, the market went *south* (an interesting euphemism: on a map, south is at the bottom and on a graph toward the bottom usually denotes a lesser quantity), the number of children playing with hula-hoops rose *exponentially* (meaning at a very rapid rate). Quantum is a very popular importation from mathematical physics: Mankind made a *quantum* leap.

I think the importation of mathematics is initiated by people who have studied some mathematics at some point in their lives. The terms stuck in their memories, and they feel that using them adds to the panache or provides scientific (and hence higher-quality) overtones to their products or enterprises. Call this invasion of mathematical terms into every-day English *mathglish*. The general public is very likely not aware of the mathematical origin of the term, but this displays yet another way that math education enters into our lives. One might point out, with reason, that these words were originally pulled into mathematics from other languages (often Latin or Greek) and from other contexts. No matter.

All this is trivial stuff, but there are invasions of mathematics into everyday life that are more important. The importation of mathematics into baseball strategies is one of them. Of the major American sports, baseball, with its relatively simple, static, and discrete structure, is surely the most mathematized. This has been the case for a very long time: Batting averages go back to 1874. Other sports, such as basket-ball, hockey, and soccer, have a continuous flow in time and space and are therefore more difficult to model. For some fans, "stats" are the name of the game, more important, even, than observing the play on the field. Baseball, whatever else it is, is also a game of chance, and chance plays a greater role in the outcomes than the "pure" skill of the players operating deterministically would lead us to believe.

For the mathematically minded and probability sophisticated, chance has a great effect on the observed patterns of wins and losses in baseball. Therefore, baseball might be modeled as a discrete Markoff chain that has many internal states and whose matrix of transitional probabilities linking the states has been estimated by passionate data collectors. This possibility was reflected in the baseball board games of years ago that had their spinners and select-a-card piles that divided non-uniformly to reflect various probabilities.

Numerous models, named after their creators Palmer, Lindsey, Mills, James, and others, are now available for the dugout strategists to use. The models take into account such things as home vs. away, grass vs. turf, players being "hot" or "cold," measures of offensive performance, and measures of performance at moments of crisis.

Baseball caps, buttons, and T-shirts bearing teams' logos have been available for years. What also are here these days are on-line and CD-ROM models, simulations of whole seasons, of All-Star and World Series Games, determinations of optimal lineups, projections of future performances, and even on-line betting capabilities. All these features are based on vast databases from which at least fifty different kinds of statistical parameters—including at bats, ratios of ground outs to fly outs, and outfield assists—have been culled.

These developments are not meant just for TV fans. Long years of managerial expertise and common sense are being replaced by stats, models, and simulations. Hirings, dismissals, and salaries are affected. Ball players now live under the tension and uncertainties created by the mathematizations of the game they play.

With models now competing against models, baseball is a vastly different game than the one familiar to Abner Doubleday in 1845. According to a well-known anecdote, Bill Klem (1874–1951), baseball umpire, hesitated for a moment before calling a crucial pitch. The batter turned around and demanded, "Well, was it a ball or a strike?" Klem responded, "What do you mean was? Sonny, it ain't nothing at all until I call it!" I assume that umpires' indices of subjectivity have been factored into today's models.

Further Reading

Jim Albert and Jay Bennett. *Curve Ball: Statistics and the Role of Chance in the Game*. Copernicus, 2001.

Philip J. Davis. "Playing Ball Probabilistically." *SIAM News*, Vol. 35, No. 3, 2001.

Mickey Flies the Stealth: Mathematics, War, and Entertainment

Mathematics, war, and entertainment—what a strange trio! The mixture of fun and glory, brutality and suffering that inheres in its story seems to contravene common sense and would require an Aeschylus to get the proportions right.

In July 2002 I attended a large conference of the Society for Industrial and Applied Mathematics, an international organization, where the range of topics discussed by the conferees was enormous, from fluid dynamics to biomathematical problems to econometrics to computer vision to mathematical cryptography. A major talk by Christoph Bregler on computer animation of human movements including facial expressions feeds right into the needs of this strange trio.

In talking to the conference participants, I learned that the "marriage" of war, mathematics, and entertainment is now taken for granted, just as the marriage of entertainment, mathematics, and medical imaging is taken for granted. The link in both cases: computer graphics, animation, and the formation of virtual objects.

The threefold combination first entered my consciousness forcibly when I read a 1997 report of the National Research Council entitled *Modeling and Simulation: Linking Entertainment and Defense.* This report was based on a workshop held in October 1996 that was attended by people from the film, video game, and theme-park industries and also by people from the defense department and their contractors. The object of the conference was to discuss common interests and how they might help each other.

Since the separate concepts of entertainment, mathematics, and war, as well as their ambiguities and misapprehensions, are sufficiently understood, I will build up to the three-fold combination in stages by beginning with three two-fold engagements.

The combination of mathematics and entertainment is ancient. Simple puzzles or games of chance and of strategy often have a mathematical underlay. More recently, sports of all kinds have experienced increasing mathematization through the accumulation of statistics and the decisions based on them. Mathematics itself is often considered a game. In his introduction to Johan Huizinga's famous *Homo Ludens: A Study of the Play Element in Culture,* George Steiner wrote "Of all human activities ... pure mathematics comes closest to Huizinga's own standards of elevated play."

In education, mathematics is often promoted on the basis that mathematics can be "fun." Computer graphics, which, as we shall see, has a substantial mathematical base, is used for the simulation of humans and environments, both static and animated, and is combined with very elaborate and sometimes interactive story lines.

The association of mathematics and war is very old, going back at the very least to Archimedes (220 BC). Niccolo Tartaglia (*La Nova Scientia,* 1537) wrote on the mathematical science of ballistics and warned that such knowledge, if widely known, could be dangerous. Since the beginning of the 20th century, war has become increasingly mathematized in all its aspects, and research in pure and applied mathematics has been supported in a great measure out of governmental offense and defense budgets.

The combination of war and entertainment is now commonplace; today's computer games are mostly simulated battles. Though virtual war flourishes via the computer, it did not require the computer to link war and entertainment. Homer was one of the first to provide the mix: Think of the gladiators in the Roman arenas or the mock naval battles in Versailles that amused the French Court. Comic books are filled with war situations. Live reenactments of old battles flourish. I recently saw on TV a reenactment in full costume of a 10th-century battle in Ireland between the Irish and the Vikings. Spectators were watching the engagements from the sidelines while the children ate ice cream.

> Ever since words existed for fighting and playing, men have been wont to call war a game.
> —Johan Huizinga

Combining all three elements, we have entertainment, mathematics, and war. Although it would be easy to give simple examples from antiquity (chess, for example), I'll limit my discussion to recent instances, where the possibility of applications to armed conflict is the principle motivation. The link I want to stress is simulated action and environments achieved through computer graphics. On the entertainment side, computer graphics appear increasingly in Hollywood productions. In the movie *The Perfect Wave* computer simulated waves appear and the technique is used to create online virtual surfing on beaches.

Mathematical theories and constructions (algorithms) combined with incredible developments in computer hardware create a solid base for computer graphics, which are increasingly used in military instruction and strategic matters.

We may also note that the principal areas of overlap between entertainment and defense are, according to the 1997 report of the National Research Council,

1. Virtual reality technologies; the creation of simulated immersive environments against which defensive–offensive action takes place;

2. Rapid communication between many—possibly thousands—of simultaneous players (i.e., combatants);

3. Computer-generated characters "that model human behavior in activities such as flying an aircraft, driving a tank, or commanding a battalion such that the participants cannot tell the difference between a human-controlled force and a computer-controlled force."

There are, of course, great differences between entertainment and the military culture. The goal of the entertainment industry is to create products that capture the public's imagination, to hook the public on such products, and to make money. The goal of the military is to defend the country and to win battles. A 2004 release from the MOVES (MOdeling, Virtual Environments and Simulation) Institute at the Naval Postgraduate School states that two of its focus areas are "computer-generated autonomy and computational cognition" and "combat modeling and analysis." These concern themselves with how to model human and organizational behavior in simulation systems as in, e.g., recent terrorist behavior and asymmetrical warfare. According to Michael Capps, Perry McDowell, and Michael Zyda of the Naval Postgraduate School, the final products of both entertainment and war are "ultimately judged on immersiveness, ease of use and realism."

It would appear that with the investment of billions of dollars in computerized filmography and video games, the entertainment industry has been ahead of the military in all three of the above areas of mutual interest despite the computer sophistication of the military. The Internet was initially a military communication arrangement that rapidly exploded into worldwide use and misuse. The military therefore can communicate its simulation needs to the entertainment world and subcontract for the latter's specialized products—a very good thing for sales. Reciprocally, assuming freedom of communication, the computer entertainment world can benefit from the techniques developed independently by the military. Simulated warfare can be bought or downloaded from the web and played. According to Capps, McDowell, and Zyda, "the Navy recently began training prospective student aviators using a commercial off-the-shelf flight simulator and found that it improves their performance."

The principal computer games have scripts that set up conflicts between the human player or numerous other players and the computer. The rules of engagement can be made as complex as desired. Thus, a description (you can find it on the Web) of a game called *Mech Commander 2* says, in part,

> On the full 3D (three-dimensional) battlefield, you will control movement, targeting, and engagement tactics of your mechs [i.e., equipment]. Call in support elements such as artillery or air strikes, scout choppers and sensor probes, even a salvage team to salvage downed mechs. Capture resource buildings or resource trucks for more "support points"... It's all up to you.

The military can outline its own specialized scenarios for training or logistical purposes, such as dealing with an ambush or freeing hostages. In order to make a forceful training impact, the virtual environment against which the scenarios are played must be tremendously realistic. Game producers are now talking about "haptic" applications (i.e., pertaining to the sense of touch) that go beyond sight and sound by wiring the players. Even without such advanced technology, cyberartists have been able to produce violent reactions such as nausea or convulsions in viewers.

The achievement of realism has a strong mathematical base. While the general goals of the trio of war, mathematics, and entertainment are in the open literature, the mathematical or programmatic details of products are either company or militarily confidential.

Modeling. The characters in computer games (people, equipment) are built from graphical prototypes, such as spheres and cylinders. Other primitives are built up from meshes and mathematical spline curves, the latter having been developed during World War II and saw early employment in the design of airplane, ship, and automobile surfaces. I understand from talking to mathematical technologists in the auto industry, that this industry now does a good fraction of its design work using graphic packages and immersion into virtual reality constructions.

Articulation and animation. Motion through space is guided by mathematically defined spline curves. Differential calculus applied to spline curves can, for verisimilitude, determine how to compute the direction of the tires in a moving truck. Martha Gregg and Paul Davis note how a realistic image of the surface of a waving flag is made: "A sine wave plus noise is imposed on the vectors normal to the triangular mesh elements defining the surface."

Shading and reflections. Shading and reflections use textured micropolygons and surface normals. Capps, McDowell, and Zyda point out that producers are claiming 66 million textured polygons per second, whereas 4.8 billion are required to make computer images absolutely indistinguishable from reality. It is predicted that advances in computer hardware will reach this goal shortly. (I would like to add that my field of mathematics is known as "approximation theory." Approximation theory concerns itself with the representations of arbitrary curves or data through mathematical formulas. By training students in the techniques of this field, I had a (small) finger in the early development of CAGD (computer aided geometric design). At the time I could hardly have foreseen the widespread uses of this mathematical technology.)

The triangle is now closed: mathematics to entertainment to war to mathematics again. The products of these links can be employed for military training on the ground, in the air, or on the sea, or for the study of strategic and tactical alternatives. For recreational purposes, the links can provide history buffs or partisans with the ability to sit in front of their terminals and, by merely clicking, refight the Peloponnesian War, the Battle of Lepanto or the capture of Paschendaele (a half-million lives lost in three months), all made realistic by mathematical modeling. Virtual casualties in their last agonies can be strewn about a three-dimensional landscape by random number generators. The line between real and virtual war becomes more and more blurred. The Great Computer Game becomes a military training ground or a Grand Guignol Theatre with audience participation.

Can the virtual ever replace the real? If simulated and networked military engagements, played over and over again, resulted in the virtual destruction of most players, then some small measure of sense might seep into the collective brain of humanity. Could we achieve through mathematical simulation what the humane philosopher William James sought for and despaired of finding: a moral equivalent of war. Could we achieve the dream of Leibniz that if a dispute arose we would standardly say "Let us compute." This would indeed be the glory of mathematics.

Further Reading

Michael Capps, Perry McDowell, and Michael Zyda. "A Future for Entertainment-Defense Collaboration." *IEEE Computer Graphics & Applications*, January/February 2001.

James Der Derian. *Virtuous War: Mapping the Military-Industrial-Media-Entertainment Network*. Westview Press, 2001.

John Leland. "The Gamer as Artiste." *New York Times*, Week in Review, Sunday, December 4, 2005.

Michael Zyda. "The Naval Postgraduate School MOVES PROGRAM—Entertainment Research Directions." In *Proceedings of the Summer Computer Simulation Conference, Vancouver, 16–20 July 2000*. SCS Press, 2000.

Michael Zyda, *MOVES 2004: Where We Came From, Where We're Going*, Publication of The MOVES Institute, Naval Postgraduate School.

Michael Zyda and Jerry Sheehan, eds. *Modeling and Simulation: Linking Entertainment & Defense*. National Academy Press, September 1997.

For more on MOVES, see also the following website:
http://www.movesinstitute.org

For seventeenth-century title page engravings in mathematical books that depict the "legitimization" of mathematics and war, see the following reference.

Volker Remmert. *Widmung, Welterklärung und Wissenschaftslegitimisierung: Titelbilder und ihre Funktionen in der Wissenschaftlichen Revolution.* Harrassowitz Verlag, 2005.

The Media and Mathematics
Look at Each Other

> Whatever has been learned about how to get at the curve of someone
> else's experience and convey at least something of it to those whose own
> [experience] bends quite differently, has not led to much in the way of
> bringing into intersubjective connection [these two people.]
>
> —Clifford Geertz, anthropologist

The treatment that mathematics receives in the media is at once a reflection and a reinforcement of common sense views. In this essay, the word *media* will mean not only the newspapers and television, but also novels, stories, plays, movies, museums—in short, all modes of popular communication of that might shed light on mathematics.

First, a few selected a few selected views about mathematicians.

1. Jonathan Weiner, a Pulitzer Prize-winning science writer, sent me a personal anecdote: Some years ago, he and his wife took their two boys out for a meal at one of those family restaurants where the tables are covered with paper tablecloths and there are crayons for the kids. While they waited for their orders, the boys entertained themselves by doing mathematics together—scribbling equations on the tablecloth. This impressed the waitress; she stared at the boys slack-jawed. Then she asked the older boy what he wanted to be when he grew up. "A writer," he said. The waitress's jaw dropped even lower. "You're so smart," she said, "and you want to be a writer?"

2. Go back three centuries and ask what Jonathan Swift's view of mathematics was. In *Gulliver's Travels* (1726) he mocks the Laputian mathematicians:

    And although they are dexterous enough upon a piece of paper in the management of the rule, the pencil, and the divider, yet in the common actions and behavior of life, I have not seen a more clumsy, awkward, and unhandy people, nor so slow and perplexed in their conceptions upon all other subjects, except those of mathematics and music. They are very bad reasoners, and vehemently given to opposition, unless when they happen to be of the right opinion, which is seldom their case. Imagination, fancy, and invention, they are wholly strangers to, nor have any words in their language by which those ideas can be expressed; the whole compass of their thoughts and mind being shut up within the two aforementioned sciences.

3. Fast-forward to Arthur Conan Doyle's *The Final Problem* (1893), where the mathematical genius is a criminal:

> [The arch-criminal Col. James Moriarty] is a man of good birth and excellent education, endowed by nature with a phenomenal mathematical faculty. At the age of twenty-one he wrote a treatise upon the binomial theorem, which has had a European vogue. On the strength of it he won the mathematical chair at one of our smaller universities, and had, to all appearances, a most brilliant career before him.

A brilliant career in the mid 1800s on the strength of the binomial theorem? Does this indicate ignorance, naiveté or satire on the part of Conan Doyle? One might rescue him from ignorance by saying that Moriarty, in advance of his time, worked on the summability of divergent binomial expansions.

The Complaints of Mathematicians

Briefly, mathematics gets very little coverage of recent developments in the papers; it would like much more. And when mathematics is covered by the papers, the papers frequently get it wrong. Technical details are mostly omitted, and their "take" or point of view is irksome. Years ago, Ronald Rivlin, an applied mathematician and a world authority on rheology (the study of the deformation and the flow of matter), was invited by *Scientific American* to write an article on his specialty. He did so, but his article was edited in a way that displeased him. A controversy with the magazine editor ensued, and the upshot was that Rivlin asked that his name be taken off the article.

Dr. James Crowley, CEO of SIAM (Society for Industrial and Applied Mathematics) wrote me,

Media professionals (writers, reporters) will often tell you that what they seek are items that have "human interest" or that deal with controversial issues. Unfortunately, for mathematics, "human interest" usually involves a stereotypically nerdy or flaky representation of a mathematician. And controversy [which is stock-in-trade of the papers] is hard to find within mathematics. We tend not to make statements of global warming or stem cell research, but rather whether a given theorem is true or not—hardly the kind of thing to generate heated debates among people on the street.

My main complaint about the treatment of mathematics in the media is that mathematics is not treated in the media. One need only look at the *Science Times* on Tuesdays to verify that there is little covered in our area of research. I think this is a general problem, though, and not entirely the fault of the media itself. Look at the press releases issued by the National Science Foundation. One sees very little having to do directly with mathematics or mathematicians.

In 1994, Fields Medalist William Thurston wrote a well-received article in the *Bulletin of the American Mathematical Society.* Here are a few passages.

We mathematicians need to put far greater effort into communicating mathematical ideas. To accomplish this, we need to pay more attention to communicating not just our definitions, theorems, and proofs, but also our ways of thinking. We need to appreciate the value of different ways of thinking about the same mathematical structure.

We need to focus far more energy on understanding and explaining the basic mental infrastructure of mathematics... This entails developing mathematical language that is effective for the radical purpose of conveying ideas to people who don't already know them.

Displeasure with the media and its effect on public understanding extends also to scientific reportage. Harvard Professor of Physics Lisa Randall writes,

It would be better if scientists were more open about the mathematical significance of their results and if the public didn't treat mathematics as quite so scary... A better understanding of the mathematical significance of results and less insistence on a simple story would help to clarify many scientific discussions.

Easier said than done. A review in the non-technical *New Yorker* magazine of Randall's newly appeared *Warped Passages: Unraveling*

the Mysteries of the Universe's Hidden Dimensions implied that the
details of the physics weren't getting through, and the mathematics was
relegated to an appendix.

A reviewer in the *Scientific American* of *The Best American Science
Writing, 2004* writes,

> Biology, physics, biotechnology, and astronomy, to anthropology, genet-
> ics, evolutionary theory, and cognition, represent the full spectrum of sci-
> entific writing from America's most prominent science authors, proving
> once again that "good science writing is evidently plentiful".

Where is mathematics in this list of the best science writing?

The Media's Point of View

One day a local ABC-6 TV reporter came around to the offices of the
American Mathematical Society (AMS) in Providence, RI. He looked
at the display in the lobby and then spoke to a staff member of the
Society who does "public awareness." He said to her, "I've passed this
building many times and I've often wondered what on earth goes on
inside. But I really don't want to know." Nothing came of his visit.

Gina Kolata, a widely-read science writer for the New York Times,
wrote me:

> You have to ask what newspapers are trying to do. There are hundreds
> of articles every day crying out for readers' attention. And every article
> has to tell a reader: "why am I reading this and why am I reading this
> now?" The News can feature something incredible, like the solution to
> Fermat's last theorem or the cloning of a dog, or it can involve some issue
> that has a big impact on society like health care, or it can be a discovery
> of something quirky. Newspapers are not there to educate or to teach
> people about the mathematics that underlies search engines unless there
> is something you can say about that mathematics that makes it new and
> compelling. The fact that the mathematics is there is not enough. With
> most things we use—a car, an iPod, a DVD, most of us don't really care
> how it works.

Sara Robinson, a science writer who majored in mathematics and
was Writer in Residence at MSRI (Mathematical Sciences Research
Institute, Berkeley, California), recalled that, as a fledgling reporter, she
heard from a senior reporter that the goal of science reportage was to

give the reader merely "an illusion of understanding of the technical subject matter." She also reported a statement made by Rob Finer, a former editor of the *New York Times*, that "Mathematics has no emotional impact. What physicists do challenges people's notion of origins and creations. Mathematics doesn't change any fundamental beliefs or what it means to be human." In view of the fact that the increasing mathematization of life is changing what it means to be human, this strikes me as a completely imperceptive view of the matter.

Novels, Plays, and Movies

In the last two decades, novels, plays, and movies with a mathematical basis have come thick and fast. In *Morte Di un Matematico Napoletano* (Death of a Neapolitan Mathematician, 1992), a brilliant, eccentric, alcoholic mathematician lives in increasing isolation and commits suicide. In *Good Will Hunting* (1997) Will Hunting, abused as a child, living a rough life in South Boston, and employed as a janitor at MIT, is discovered to be a math genius by a Fields Medal–winning professor; with the aid of a therapist, his life is then turned around. In *Arcadia* (1999) by Tom Stoppard, a teenage genius living several hundred years ago discovers that entropy of the universe is increasing.

More recently, there has been much discussion about Fermat's Last Theorem, chaos, determinism à la Newton and Laplace, population dynamics, etc. The musical *Fermat's Last Tango* (2000) won an Emmy Award; in fantasy, the spirits of Fermat and other mathematical greats residing in a heavenly *Jenseits*, or Aftermath, meet up with an obsessed mathematical hero reminiscent of Andrew Wiles of Fermat fame. In *Uncle Petros and Goldbach's Conjecture* by Apostolos Doxiadis, a mathematician and movie maker (English translation, 2000). The hero is obsessed by his inability solve a difficult problem; giving up, he retires in defeat to playing chess and tending his garden.

The movie *A Beautiful Mind* (2001) won four Academy Awards and was based vaguely on the biography of mathematician John Nash. A great mathematician descends into madness and then makes a partial recovery. The movie's emphasis is on the mental state of the mathematician and not on the particular mathematics achieved. The play *Proof* won a Pulitzer Prize. The movie (2005) starred Gwyneth Paltrow and

Anthony Hopkins. A deceased, mentally ill mathematician had a daughter who devoted her life to caring for him. His daughter and a former student of the mathematician come across the proof of an important theorem in the mathematician's residual papers. The question of to whom the proof is due is raised, as is the question of whether the daughter has inherited her father's madness.

The impression gained from these stories is that mathematicians are strange and peculiar people.

A Database of Mathematical Fiction

Alex Kasman of the College of Charleston, South Carolina, is a mathematician, author, and archivist for literary works with a mathematical underlay. Kasman maintains a website titled *Mathematical Fiction* that contains brief summaries and opinions of about 500 works. I would not have guessed that there have been so many works of this sort, and I am indebted to his website for some of my information.

Kasman has divided and cross-referenced these works into about 50 categories. Under the rubric of "Medium" we have comic books, films, novels, plays, short stories, TV series or episodes). Under "Genre": children's literature, didactic, historical fiction, humor, espionage, fantasy, horror, mystery, and science fiction. The writers of these works range from satirists such as Jonathan Swift and Stanislaw Lem to novelists like Robert Musil, who studied mathematics as a young man. They range from science-fiction writers such as Jules Verne, playwrights such as Pirandello and Tom Stoppard (who is a mathematics buff), and humorists such as James Thurber. They include professional mathematicians such as Charles Dodgson and my doctoral thesis advisor Ralph Boas, Jr.

Kasman's list contains a wide variety of themes and types of work written by people with different degrees of professional mathematical knowledge and experience. The number of professional mathematicians who write fiction is greater than I imagined. While the whole genre is not easily characterized, I would call about half of the individual works catalogued "sensationalist."

For example, in *The Rose Acacia*, by Ralph P. Boas, Jr., a computer made a deal with the Devil as to exactly how many terms of a very slowly convergent infinite series it takes to arrive at two accurate places of the sum.

In *Pop Quiz*, by Alex Kasman, messages from an alien spacecraft seem to be asking deep questions in algebraic geometry. What is the intent of all the messages?

What is clear to me is that there is much material that supports the assertion that a mathematician is brilliant, somewhat mad, socially inept or reclusive, obsessive, living in the clouds, given to the arcane or fantastic. The mathematicians or mathematics depicted come wrapped in the following sensational themes: magic, codes, espionage, the devil, ghosts, secret messages, other worlds, futurism, madness, autism, apocalyptism, mysticism, the occult, obsessions, prizes, distopias, evil mathematical productions and cults, machines that turn into sorcerers' apprentices, alternate time concepts. Apparently there is a steady market for this kind of literature, and mathematicians themselves are writers, producers, and readers of it. The average person often regards mathematics as a kind of magic, and this view fits right in with the fictional themes. Some professional mathematicians like to write this kind of material, and Pulitzer Prize-winning *L.A. Times* writer Dan Neil was aware of all this when he wrote "Never have so many relied on so few to tell us what the hell is going on. Mathematicians have acquired the status of hieratic otherness, a kind of geek priesthood, acting as intermediaries between the unfathomable and familiar." [Definition: geek = an obsessive specialist]

Of course, there is another way of regarding these productions. A leading research mathematician told me that he factors out the mathematical descriptions, which in any case he finds trivial, and concentrates on the story line, for instance, the relationship between a man and his nephew in the novel *Uncle Petros*.

The average person has had little hands-on experience with mathematics other than, perhaps, doing a few sums and paying interest on a variety of purchases. The common-sense view of mathematics and mathematicians—or what masquerades as common sense—is to a considerable extent formed by what the public remembers from grade school or what they now learn from the media. "You do the math" is now a common expression that epitomizes distance from the whole magic,

incomprehensible, tedious enterprise. It may be that with its abhorrence
of mathematics, the general public ignores the mathematics totally and
simply goes for a whopping good story. I would suggest a substantial
financial grant to an appropriate university department for a survey to
test this statement.

Is there any reality to these stereotypes?

I believe there is some, and it starts from the simple observation that
some people are good at mathematics and some are not. Some people
care about mathematics deeply, they think about it constantly. Other
people couldn't care less. Nature and nurture? Presumably, and on the
side of nature, common observation leads to the speculation that some
people are hard wired for mathematics (whatever that expression may
mean).

Chapter 6 ("Are People Hard Wired to Do Mathematics?") contains a
discussion of the possible association of extreme brilliance with Asperger's Syndrome, and several candidates for Asperger's were mentioned.
Be that as it may, I'm sure that most professionals have had contact
with some mathematician that they would put in this category. And as
Clifford Geertz wrote, "The rational beauties of mathematical proof are
guarantees of no mathematician's sanity."

Where Does the Difficulty of Communication Lie and What Can Be Done?

There is a paradoxical disconnect between the substantial mathematization of everyday life to which everyone is subjected and the extreme
reluctance of the general public to learn anything of the subject beyond
grade school material. Why should they know more when the relevant
numbers and implications are spewed out automatically? Many believe,
like television star Rosie O'Donnell, "There's no way they should have
to teach it [mathematics], now. We have computers."

Where does the difficulty lie? Does it lie with mathematics education
in the lower grades, often taught by teachers for whom the subject is
poison? Doesn't every specialized activity, from cooking to dentistry,
have specialized terms that the public becomes familiar with? Shouldn't
grades K–12 provide a basic vocabulary and infrastructure of understanding so that the average newspaper reader will not turn the page

rapidly upon encountering the word "mathematics"? Does the difficulty lie with the nature of mathematics, which, at its higher reaches, is a difficult subject requiring much more time for absorption and patience on the part of readers than newspaper articles do? Does the difficulty lie therefore with the nature of the media with its promise of instant information and understanding? Does it lie with the journalists who write about recent accomplishments in the field? with the mathematicians themselves who, in the opinion of Dan Neil, constitute a geek priesthood? It would appear that all of the above are operative.

A west coast mathematical sciences writer (with a BA in the subject) said to me recently that there are now many books written by professionals to popularize their specialties. "And you know what? After five pages, they've lost me."

James Crowley of SIAM points out that

> The reason that we tend not to report on mathematics may have something to do with the nature of the discipline itself as well. Because most discoveries in mathematics are incremental, building upon a vast and growing structure of knowledge, there is seldom one breakthrough discovery that can be announced with one big "aha!"

A number of programs intended to cure the situation are in place. The professional mathematical societies and a number of individual writers work hard to call the recent accomplishments in the field to the attention of a variety of clienteles at every level of sophistication. Thus, the American Mathematical Society has a number of items devoted to this task. It publishes a *Math Digest* that summarizes popularizing articles and articles about mathematics in the popular press. It produces *Mathematical Moments*, a series of pamphlets that "[promote] the appreciation of the role of mathematics in science, nature, technology, and human culture" to schools and colleges through flyers, suitable for display, on such individual topics as forecasting the weather, creating better eyeglass lenses, etc. What to a great extent gets released to newspapers is information such as the names of recent prize winners, adding to the absolute glut of prize winners of all sorts in today's world.

The Division of Mathematical Sciences at the National Science Foundation is trying to gather information about mathematics and its applications to use in press releases. This is not an easy job, because few people in the mathematical sciences are accustomed to reporting on their work at a level that can be appreciated by a general audience; it is not part of their culture. Some institutes, such as MSRI in Berkeley, have a rotating journalist in residence. The United States Air Force, in response to fewer and fewer native American students pursuing science, technology, and mathematics, has given a grant to the University of Southern California to teach screenwriting to scientists in an attempt to produce movies and television shows that show scientists in more sympathetic ways than is currently done. How all these professional projects diffuse into the popular media and public awareness is a matter of conjecture.

What Kind of Reportage Would I Like to See?

As an antidote to sensational reportage, I would suggest that newspapers run articles that give a semblance of understanding of the degree to which mathematics underlies today's world. We are living in an increasingly mathematized age. To realize this, take a look at the front page of a newspaper and count the numbers that are on it. Some numbers invoke trivial mathematics; some less so. Go to the business section and do the same. Go to the sports section and do the same. The mathematics involved is often of long standing and is not material hot off the research pages.

Just think what a different (perhaps better) life we would lead if IQs were not around to tell us who is "intelligent" and who is not. How would we live if there were no blood pressure or cholesterol numbers to advise us? if there were no pre-election polling on every conceivable issue? if the trajectory of a missile or of a rocket to Mars were computed by pre-Newtonian theories or by guesswork? if, when we went to the supermarket, the checkout clerk took a pencil from behind his ear, marked down the prices of our 18 items and toted them up? Do you know how the House of Representatives is apportioned among the States after a census? It's done via a mathematical procedure known as the "method of equal proportions" that is part of statutory law (Title 2, U.S. Code) and judged constitutional by the U.S. Supreme Court. Have you looked at a weather map with its isobars and isotherms? There are mathematical algorithms that underlie the production of these pictures.

What meaning can be ascribed to the statement that "there's 40% chance of rain tomorrow"? Did you know that diffusion equations have been invoked to advise orchid growers how often to water their plants?

These experiences, and hundreds more, have been created through an underlay of mathematical ideas and methods, some trivial, some deep and by no means within the abilities of the general public, and yet few of them have been called to the attention of the public as something worthy of its attention. Coming close to expositions of these mathematical ideas are the periodic releases entitled *Mathematical Moments,* mentioned above, but the average reader cannot handle their vocabulary.

In contradiction to what Kolata has said, why is it not the duty of a newspaper to educate? Doesn't a paper educate when it admonishes us not to do drugs? when it advises us how to make a tofu soufflé? when it reminds us that there is a congressional election every two years? or that it is time to push the clock back one hour? I should think that at the very least it should be possible for a paper to educate us to the fact that mathematics is formatting a good portion of today's life and to point out where this is occurring. It need not give the readers a semblance of understanding of the technical mathematics; that is too much to expect. But I should hope that clever writers might point out how mathematics is altering our lifestyles, and do it in a manner that would not lead Garfield the Cat to say "ho hum."

Is it too much to hope that my local newspaper might in the future run articles with titles such as

> Professor Hiram Smith Shows How Eigenvalues Help in Search Engine Strategy; President Asks Science Advisor What Egg Values Are

> Numerical Algorithms of Aero-hydro-elastodynamics Used in the Design of the Swiss Yacht that Won the America's Cup

> Algorithms for Inverse Problems in Differential Equations Aid in the Evaluation of MRI Data at Beth Israel Hospital

> The Surgeon General Urges Medical Schools and the Attorney General Urges Law Schools to Require Probability Theory for Admissions

> Yogi Berra Praises Markoffian Applications to Baseball Strategy

I hope that such items have already been run and that as I read my morning paper with my eggs and coffee I simply missed them.

Postscript

This essay represents my view of the American scene *vis-à-vis* mathematics and the media. My correspondents in Europe tell me that with some slight modifications it describes the European scene equally well.

Further Reading

Philip J. Davis. "Mathematical Exhibitions: Reactions and Concerns." *SIAM News*, Vol. 38, No. 10, December, 2005.

J. Dieudonné. "Genius and Biographers: The Fictionalization of Evariste Galois." *American Mathematical Monthly*, Vol. 89, No. 2, 1982.

Paul W. Droll. "Orchid Watering Decisions." *Orchids*, Oct. 2005.

Clifford Geertz. *Local Knowledge*. Basic Books, 1983.

Erica Klarreich. "Mathematics in the News." *Emissary* (Newsletter of MSRI), Spring 2002.

Dan Neil. "Who Knew Math Was So Prime Time." *Los Angeles Times Magazine*, October 9, 2005.

Rosie O'Donnell Show, March 23, 2001.

Lisa Randall, Op Ed *New York Times*, September 18, 2005.

Sara Robinson. "Mathematics and the Media: A Disconnect and a Few Fixes." *SIAM News*, October 2001.

Rina Zazkis and Peter Liljedahl. *Hollywood Perceptions of Mathematics: Cultural Truth or Mathematical Fiction?* Public lecture presented at Faculty of Education, Simon Frazer University, 2003.

See also the following websites:
http://www.ams.org/mathmedia/mathdigest/
http://www.ams.org/ams/dbis.html

Platonism vs. Social Constructivism

It is written that animals are divided into (a) those that belong to the Emperor, (b) embalmed ones, (c) those that are trained, (d) suckling pigs, (e) mermaids, (f) fabulous ones, (g) stray dogs, (h) those that are included in this classification, (i) those that tremble as if they were mad, (j) innumerable ones, (k) those drawn with a very fine camel's hair brush, (l) others, (m) those that have just broken a flower vase, (n) those that resemble flies from a distance.

—Jorge Luis Borges, *The Analytical Language of John Wilkins*, In: *Other Inquisitions*

What is the nature of the vast body of concepts, experiences, perceptions, formulations, practices, computations, and applications called "mathematics"? What is the relationship between the people who are engaged in mathematics and the mathematics itself? I've singled out mathematics, but the same questions can be asked of any branch of science. Mathematics and philosophy have had a long love affair. Some scholars have said that philosophy is a Greek invention brought on because of the Pythagorean paradox that $\sqrt{2}$ both exists and doesn't exist. Be that as it may, every time critics raise their eyebrows about the sense of a new piece of mathematics, a new chapter of the philosophy of mathematics is opened. Most often, the reason for raising the eyebrows was that the new theories seemed to go against common sense.

The philosophy of mathematics accommodates itself to whatever mathematicians create, discover, do, or talk about. Ludwig Wittgenstein pointed out correctly that "even 500 years ago, a philosophy of mathematics was possible, a philosophy of what mathematics was then." There are now at least eight distinguishable schools of mathematical philosophy, and new developments such as the digital computer have given rise to new schools. The dominant philosophy of the age is taken to be the common sense about the matter.

Most working mathematicians and scientists consider philosophic questions to be irrelevant to the daily pursuit of their respective specialties, but such questions have been around for three thousand years and have been mulled over by the finest brains. The questions are still here and have plenty of fuel left in them to arouse passions to the flash point and beyond. Sometimes the answers given influence educational policy.

I shall consider here only two schools: Platonism and social constructivism.

One view of the nature of mathematics is that mathematical truths exist independently of humans. They are universal, independent of time, and valid in all conceivable worlds, perhaps, even, in the absence of material worlds. In this view, the job of the mathematician is merely to dig out, validate, display, and use these truths. This is an extreme view, and one that has often been parodied by saying "the π is in the sky." This view is generally called Platonism.

Many have found "reality" within the world of ideal Platonic forms and conceptions. As mathematical physicist Roger Penrose writes, "[The] ancient Greek insight that it is mathematics that underscores the workings of physical reality has served us remarkably well..."

Opposed to Platonism is the view that mathematics is a social product constructed by groups of communicating people, and the universe could conceivably have gotten along—thank you very much—without one jot of mathematics. This statement is an extreme form of what is called "social constructivism," a philosophy that gained strength and popularity during the 20th century. The phrase "mathematical existentialism" has also been employed. (I've used this same phrase in a previous essay but in a totally different sense.) Among mathematicians, its adherents still constitute a small minority, since most mathematicians, if they think consciously of philosophy at all in their working lives, are Platonists.

Ian Hacking, philosopher and historian of science, has taken up the question of social constructivism in his book *The Social Construction of What*. He has formulated the arguments in favor of this philosophy in terms of three "sticking points." I suppose that by this phrase he means the constructivist positions that might stick in the throats of those who happen to believe otherwise.

First sticking point: contingency. The social constructivist believes that the subject of mathematics could have developed in a totally different way and led to a non-equivalent corpus of material. The opposite of contingency is necessity, and necessity comes with different shades of meaning. Thus physicists, Hacking says, employing mathematical tools, are not necessitarians. They are inevitabilists, believing that if successful physics has emerged from their work, it was inevitable that it emerged along the lines that it did.

Second sticking point: nominalism. The social constructivist is a nominalist who believes that "the world is so autonomous ... that it does not even have what we call structure in itself... All the structure we can conceive lies within our representations." The anti-nominalists would assert the existence of "inherent structures. I suppose that most scientists believe that the world comes with an inherent structure which is their task to discover."

Third sticking point: explanations of the stability of theories. Hacking says "The constructionist holds that explanations for the stability of scientific belief involve, at least in part, elements that are external to the professed content of the science. These elements typically include social factors, interests, networks, ... Opponents hold that whatever be the context of discovery, the explanation of stability is internal to the science itself."

Hacking also believes—as do I—that very few people will accept or reject these sticking points outright, so he allows for shades of belief. He invites his readers to rate their adherence to his three sticking points on a scale of 0 to 5, 0 for total disbelief in social constructivism, and 5 for agreement with the positions advocated. (Question: Do rating systems such as this reflect an objective reality or are they socially constructed? See the section *Measuring the Unmeasurable: The Subjective Becomes Quantifiable* in Chapter 8, "Quantification in Today's World.")

How does Hacking rate himself? On contingency, 2; on nominalism, 4; on stability, 3. On the basis of these ratings, I would call Hacking a mild constructivist. Whom does he dub the paragon or paradigm of constructivists? The most famous, or certainly the most discussed philosopher of science of our time, the man who took the word "paradigm" out of the appendix of Latin grammar books and put it into the lips of all philosophers and humanists was Thomas Kuhn (1922-1996). Hacking rates Kuhn 5, 5, 5, all the way.

My own self-rating would be: 4, 4, 4, which is that of a fairly strong constructivist. When I was younger, I believed Mozart's G-Minor Symphony was note-perfect, existing in heaven before the world was even created. Later I saw the struggles recorded on a page of a manuscript score. Thinking of this experience, I might up my rating on the contingency dimension to 4.5. When I remind myself Borges' parody, cited at the head of this essay, of the way we make categories, I might up my rating on nominalism to 4.8.

I admit, as does Hacking, that there are objections to philosophic social constructivism. Then again, there is hardly a philosophy that admits of no objections. Social constructivism can't be pushed too far. If pushed, it leads to the Foucaultian idea that the whole function of science is to justify the existing societal power structure or to the radical-feminist view offered by Margaret Wertheim that it reflects an oppressive patriarchy.

I now invite my readers to rate themselves; but before they do, let them think about where they stand on the contentious subjects of homeopathy and acupuncture (both of which were encouraged by the mathematician–philosopher Paul Feyerabend). Where do they stand on IQ testing; on whether measures of missile accuracy are an objective concept (think of recent military experience); on the existence of nano-bacteria; whether the true TRH (thyrotrophic releasing hormone) has been isolated; whether magnetic monopoles exist. Some of these topics are discussed in Hacking's book.

Mathematical readers may object to these pro-constructionist examples. Mathematics is different, they will say. The contraction mapping theorem or the axioms for a group have nothing to do with us or with the exterior world. Mathematics enjoys a unique position among the scientific disciplines. Or does it? Answer this old philosophical chestnut: Would there be any mathematics if the universe contained no sentient creatures? Or is the universe itself sentient? And would your answer to this question point to objective or to socially constructed knowledge?

Further Reading

Jorge Luis Borges. *Other Inquisitions, 1937-1952*. University of Texas Press, 1964.

Carlo Carlucci. Introduction to *Philosophic e mathematical*. Laterza, Bari, 2002. An English translation can be found in *Eighteen Unconventional Essays on the Nature of Mathematics*, Reuben Hersh, ed., Springer-Verlag, 2005.

Philip J. Davis. "Applied Mathematics as Social Contract." *Zentralblatt für Didaktik der Mathematik*, 88/I. Also in *The Mathematics Magazine*, Vol. 61, No. 3, June 1988.

Philip J. Davis and Reuben Hersh. *The Mathematical Experience*. Birkhäuser Boston, 1980.

Paul Ernest. *The Philosophy of Mathematics Education*. Falmer Press, 1991.

Paul Ernest. "Social Constructivism as a Philosophy of Mathematics." Thesis, SUNY Albany, 1998.

Paul Ernest, ed. *Mathematics, Education and Philosophy*. Falmer Press, 1994.

Paul K. Feyerabend. *Against Method: Outline of an Anarchistic Theory of Knowledge*. NLB, 1975; revised edition, Verso, 1988.

Eduard Glas. "Mathematics as Objective Knowledge and as Human Practice." In *Eighteen Unconventional Essays on the Nature of Mathematics*, Reuben Hersh, ed. Springer-Verlag, 2006. Also in *Perspectives on Mathematical Practices*, J. P. Van Bendegem and Bart Van Kerkhove, eds. Springer-Kluwer, 2006.

Ian Hacking. *The Social Construction of What*. Harvard University Press, 1999.

Reuben Hersh. *What Is Mathematics, Really?* Oxford University Press, 1996.

Reuben Hersh, ed. *Eighteen Unconventional Essays on the Nature of Mathematics*. Springer, 2006.

Roger Penrose. *The Road to Reality: A Complete Guide to the Laws of the Universe*. Knopf, 2005.

J. P. Van Bendegem and Bart Van Kerkhove, eds. *Perspectives on Mathematical Practices*. Springer-Kluwer, 2006.

Margaret Wertheim. *Pythagoras' Trousers: God, Physics and the Gender Wars*. W. W. Norton, 1997.

Mathematics at the Razor's Edge

You taught me language, and my profit on it is I know how to curse.
—Caliban, in William Shakespeare's *The Tempest*

I come finally to what is probably the most important and perplexing question relating to mathematics and common sense. Where is mathematics going, and what is the common sense of society's support of mathematical research and of its applications? What is the common sense—or the opposite—of creating more mathematics? Isn't there more than enough of it already?

Here are some answers that have been given over the centuries for why people work with mathematics, teach it, apply it, create more of it, philosophize about it, and even despise it. In each period of time, society has decided for which of these reasons it wishes to support mathematics with understanding, respect, honors, and with cold hard cash.

Technological. Mathematics is the language of science and technology. It helps create new physical theories. It helps predict the future. A good part of contemporary technology has a substantial mathematical base and carries with it the possibility of unintentional effects: A search engine is a wonderful device with many fine uses and now there is psychological evidence that it has reduced individual autonomy by overreliance on easily available information.

Economic. Mathematics is the binding language of economics and a facilitator of business and commerce.

Educational. Mathematics is one of the classical liberal arts. By stressing logical thinking, it helps develop skills in diverse areas; it enables us to live intelligently and productively in an increasingly mathematized world.

Societal. Mathematics can create techniques that can lead to the alleviation of the ills of society, and, as already noted by Archimedes (c. 225 BC) and Niccolo Tartaglia (1499–1557), it can facilitate the destruction of lives.

Medical. Mathematics can lead to the understanding of the human body and assist medical procedures. It can lead to the alleviation of pain and prolong human lives.

Personal. Mathematics is an avenue for intellectual competition. The possibility of engaging in it is there. It is fun to do and leads to aesthetic pleasure. It provides an escape from an imperfect world.

Philosophical. Mathematics is the essence of rational and objective thought and hence promotes such thought. It is worth doing for its own sake and promotes an art-for-art's-sake attitude to the subject. Over the years, mathematics has contributed mightily to philosophical concepts and positions. It leads to a deep knowledge of the cosmos and, ultimately, to a knowledge of God.

Mystical. Mathematics provides a mystical and occult interpretation of aspects of the universe.

* * *

Looking to the future, some of the immediate goals of mathematical research, described in general language, are to discover (or create) new mathematical concepts, theories, and relationships; to formulate new conjectures; to prove old and new conjectures. The specifics of mathematical programs for the future have been laid out in numerous publications, particularly at the time of the millennium (the years 2000–2001).

The experience of the past 500 years is that these goals have been fulfilled in ample measure, and, as to the future, mathematical physicists John C. Baez and James Dolan expect "the amount of mathematics produced in the 21st century to dwarf that of all the centuries that came before."

Digital computers and their interlinked networks, which are mathematical and informational engines and much more, have ushered in an era of intensified activity aimed at the mechanization of mathematics. As a corollary, some seers have said that in virtue of the tremendous facility of computer-generated graphics, mathematics will depend increasingly on visualizations and less on formulaic material. The old guard will then complain, as it has on several occasions in the past, that "this stuff isn't really mathematics," and the Young Turks will put forward a reciprocal opinion.

If mathematics is a totally formalizable discipline—a proposition that for numerous reasons I do not believe—then it can be turned over completely to computers. The computers will presumably then be sufficiently intelligent to discover—without the intervention of humans—deep mathematical phenomena. One might say then, paradoxically, that one of the successful aims of mathematical thought has been to get rid of mathematical thought.

If one interprets the word "mechanization" in a historical sense, this is a process that has been going on steadily since the creation of mathematics. We have arithmetic tables going back four or five millennia. We have "state of the art" summaries of mathematics prepared in classical antiquity (Pappus, *Collections*, c. 300). We have mechanical devices such as abaci, quipu, calculating tables and machines that are equally ancient.

Can one expect revolutions in mathematics? Certainly, but only rarely. If one defines a mathematical revolution or bombshell as a development that seriously alters the philosophy of mathematics, I can think of only four bombshells—a rather stingy assessment, I admit—in the past four centuries: Isaac Newton's *Principia* (1687), János Bolyai–Nikolai Lobachevsky's non-Euclidean geometry (1832–1840), Georg Cantor's set theory (1874), and Kurt Gödel's incompleteness theorem (1930). Figuring that the arrival of a bombshell takes perhaps a hundred years, the next one should be due around the year 2030. But contemporary mathematicians like to claim bombshells for their generation. Thus Peter Lax suggested that "Milnor's differential topology, which has altered our notion of how space hangs together," should be added to my list. David Mumford points out that during the 20th century there was a gradual and revolutionary shift toward an all-embracing abstract structuralism, associated with the German school of abstract algebra. and moving toward the work of the École Bourbaki and of Henri Cartan. "It was part of the larger cultural movement that brought us cubism and abstract art." This parallelism between art and mathematics has not yet been adequately critiqued.

What will the next bombshell consist of? I don't think that the solution of any of the current crop of unsolved mathematical problems, important as they may be, will lead to the overturn of the philosophical notions now held strongly. The total mechanization of mathematics, however—if it ever gets close to that—would be a change at the meta-level, and should rumble the philosophers quite a bit.

As Caroline Dunmore wrote, "Mathematics is conservative on the object-level and revolutionary on the meta-level." I interpret this to mean that while the objects of mathematics, numbers, functions, geometrical figures, etc., largely retain their original personas, the formal superstructures in which they are embedded can change drastically (see, e.g., Saunders Mac Lane, cited below).

∗∗∗

The promised future plethora of mathematics and mathematizations and the increasing approval and support by society leads inevitably to the question of whether mathematics is, as some claim, ethically neutral. I answer that mathematics is ethically ambiguous. Søren Kierkegaard's famous *Either/Or* described the conflict between the aesthetic and the ethical. Kierkegaard opted for the ethical. Niels Bohr wrote an open letter to the United Nations (July 7, 1950) in which he argued for free scientific discourse as a means of promoting *détente*. Was this position naïve? Was it irrelevant? Or was it prophetic in that, e.g., the Internet has altered the attitudes and interactions of populations and hence of international relations?

The razor's edge of ambiguity on which mathematics is now poised is well illustrated by the career of Lewis Fry Richardson (1881–1953), an applied mathematician, physical scientist, inventor, and sociometrist. Richardson was also a Quaker pacifist who is considered to be the father of modern mathematical war-gaming. Richardson believed that modeling warfare would lead to a diminution of antagonisms. War-gaming is now a technique employed by military strategists.

Language is a symbolic system that has raised us from the level of Caliban brutes and has accordingly transformed our lives and the same may be asserted of mathematics. It is a language that has transformed our lives for good and for bad. It has been the handmaiden not only of miraculous technology but also of new and unprecedented dimensions of human cruelty. While the ethical issues in sciences are widely recognized, the ethical issues in mathematical thinking are not adequately recognized, either within the mathematical research community or in mathematical education. So what has been our profit from it? Surely mathematics has not been a zero-sum game. Nonetheless, we must remain constantly aware of the painful truth that mathematics has the Promethean power both of life enhancement and of returning us to the condition of brutes.

Further Reading

V. Arnold, et al., eds. *Mathematics: Frontiers and Perspectives*. American Mathematical Society, 1999.

Michael Beeson. "The Mechanization of Mathematics." In *Alan Turing: Life and Legacy of a Great Thinker*, C. Teuscher, ed. Springer-Verlag, 2003. Also available from the Internet (http://www.cs.sjsu.edu/faculty/beeson/Papers/turing2.pdf).

Bernhelm Booss-Bavnbek and Jens Høyrup, eds. *Mathematics and War*. Birkhäuser, 2003.

Philip J. Davis. "War, Weather, and Mathematics." *SIAM News*, Vol. 29, No. 9, November 1996.

Philip J. Davis. "Biting the Bullet: Mathematics and War." *SIAM News*, Vol. 37, No. 2, March 2004.

Philip J. Davis. "The Power of Numerics." *SIAM News*, October, 2004.

Philip J. Davis. "Running Counter to Inert Crystallized Opinion." *SIAM News*, Vol. 37, December 2004.

Philip J. Davis. *The Prospects for Mathematics in a Multi-media Civilization*. Urania talk presented at the International Congress of Mathematicians, Berlin, 1998. Also in *Alles Mathematik*, Martin Aigner and Ehrhard Behrends, eds. Vieweg, 2000.

Caroline Dunmore. "Meta-Level Revolutions in Mathematics." In *Revolutions in Mathematics*, David Gillies, ed. Oxford University Press, 1992.

Björn Engquist and Wilfried Schmid, eds. *Mathematics Unlimited—2001 and Beyond.* Springer-Verlag, 2001.

Jeremy Gray. *János Bolyai, Non-Euclidean Geometry and the Nature of Space.* MIT Press, 2004.

Richard W. Hamming. "The Unreasonable Effectiveness of Mathematics." *American Mathematical Monthly*, Vol. 87, No. 2, February 1980.

Gerald Holton. "On the Psychology of Scientists, and Their Social Concerns." In *The Scientific Imagination: Case Studies.* Cambridge University Press, 1978.

Peter Lax. Letter to editor in *SIAM News*, Vol. 38, No. 1, Jan/Feb, 2005.

Saunders Mac Lane. *Mathematics, Form and Function.* Springer-Verlag, 1986.

David Mumford, e-communication.

Laurent Schwartz. *Laurent Schwartz: A Mathematician Grappling with his Century.* Birkhäuser, 2001.